T0280936

Partial Differential Equations

Partial Differential Equations: Topics in Fourier Analysis, Second Edition explains how to use the Fourier transform and heuristic methods to obtain significant insight into the solutions of standard PDE models. It shows how this powerful approach is valuable in getting plausible answers that can then be justified by modern analysis.

Using Fourier analysis, the text constructs explicit formulas for solving PDEs governed by canonical operators related to the Laplacian on the Euclidean space. After presenting background material, it focuses on: Second-order equations governed by the Laplacian; the Hermite operator; the twisted Laplacians; and the sub-Laplacian on the Heisenberg group.

Designed for a one-semester course, this text provides a bridge between the standard PDE course for undergraduate students in science and engineering and the PDE course for graduate students in mathematics who are pursuing a research career in analysis. Through its coverage of fundamental examples of PDEs, the book prepares students for studying more advanced topics such as pseudo-differential operators. It also helps them appreciate PDEs as beautiful structures in analysis, rather than a bunch of isolated ad-hoc techniques.

New to the Second Edition

- Complete revision of the text to correct errors, remove redundancies, and update outdated material

- Expanded references and bibliography

- New and revised exercises

- Three brand new chapters covering several topics in analysis and the twisted bi-Laplacian not explored in the first edition.

- **M. W. Wong** is a professor and former chair of the Department of Mathematics and Statistics at York University in Toronto, Canada. From 2005 to 2009, he was president of the International Society for Analysis, its Applications and Computations (ISAAC).

Partial Differential Equations

Topics in Fourier Analysis, Second Edition

M. W. Wong
York University, Canada

CRC Press
Taylor & Francis Group
Boca Raton London New York

CRC Press is an imprint of the
Taylor & Francis Group, an **informa** business

A CHAPMAN & HALL BOOK

Second edition published 2023
by CRC Press
6000 Broken Sound Parkway NW, Suite 300, Boca Raton, FL 33487-2742

and by CRC Press
4 Park Square, Milton Park, Abingdon, Oxon, OX14 4RN

© 2023 M. W. Wong

First edition published by CRC Press 2014

CRC Press is an imprint of Taylor & Francis Group, LLC

ISBN: 978-1-032-07316-3 (hbk)
ISBN: 978-1-032-07409-2 (pbk)
ISBN: 978-1-003-20678-1 (ebk)

DOI: 10.1201/9781003206781

Typeset in CMR10 font
by KnowledgeWorks Global Ltd.

Publisher's note: This book has been prepared from camera-ready copy provided by the author.

Contents

Preface

Preface to the First Edition

The motivation for writing this book comes from the desire and the need that there be a bridge between the standard undergraduate course in partial differential equations for students in science and engineering and the graduate course in partial differential equations for students in mathematics who are interested in a research career in analysis.

The literature on partial differential equations is huge. It covers a broad spectrum of topics from the very classical and applied to the very modern and pure. We have chosen to look at some linear partial differential equations in a setting that is somewhere between the very classical and the very modern. Listed in the bibliography is a very small sample of references that have shaped my vision on the subject.

The focus in this book is on the constructions of solutions of partial differential equations governed by first the Laplacian on $\mathbb{R}^n$, then the Hermite operator on $\mathbb{R}^n$, and finally the sub-Laplacian and its twisted Laplacians on the Heisenberg group $\mathbb{H}^1$. The topics are chosen not only because of their fundamental importance, but also for coerciveness in the sense that the methods are largely based on Fourier analysis. It is hoped that this book may help toward an appreciation that partial differential equations can be studied as a subject with a beautiful structure in its own right and not just a bag of isolated ad-hoc techniques.

The emphasis in the book is not on complete mathematical rigor, but rather on how the Fourier transform and some heuristics can be used effectively to obtain significant insight into the solutions of the canonical models in partial differential equations. This approach is valuable for its magnificient power in getting plausible answers, which can then be justified by the norm of modern analysis.

The first part of the book consists of 13 chapters. The first chapter is on the notation currently used in the practice of partial differential equations. Chapters 2–5 are on the gamma function, convolutions, Fourier analysis, and distribution theory, which provide the background material and the *lingua franca* of the book. Chapters 6–13 are on second-order equations that bear on the Laplacian on $\mathbb{R}^n$. The real study of partial differential equations begins in Chapter 6 in which the heat kernel is constructed. The construction of the free propagator given in Chapter 7 is similar to that of the heat kernel, but the technical details are more demanding. Chapter 8 on the constructions of the Newtonian potential is closely related to the heat kernel in the sense that the Newtonian potential is the integral from zero to infinity with respect to time of the heat kernel. The Poisson kernel is constructed in Chapter 11. Chapter 13 is devoted to the constructions of the wave kernels. The Bessel potential for the perturbed Laplacian, which is the analog of the Newtonian potential for the Laplacian, is constructed in two ways in Chapter 9. The Bessel–Poisson kernel for the heat equation when the Laplacian is replaced by the perturbed Laplacian is given in Chapter 12. It is clear from this overview that standard undergraduate courses in real analysis and complex analysis are sufficient for a good understanding of Chapters 1–13.

The second part of the book contains Chapters 14–16. These three chapters are devoted to the Hermite equation, which is a prototype of a partial differential equation with variable coefficients. The Hermite operator is the simplest and yet very important operator in quantum physics. This operator and the corresponding equation are studied in some detail after a recall of the basics on Hermite functions in Chapter 14. This module requires some knowledge of Hilbert spaces up to the basic properties of orthonormal bases, which are normally taught in a fourth-year undergraduate course in real analysis. Nothing more than the Fourier series expansions in Hilbert spaces is assumed and any one of the books [12, 34, 37, 57], among others, contains enough information on Hilbert spaces for a complete understanding of the book.

Chapters 17–23 in the last part of the book contain a detailed study of the sub-Laplacian on the Heisenberg group. For the sake of simplicity in notation, we are content with the complete analysis for the one-dimensional Heisenberg group of which the underlying manifold is $\mathbb{R}^3$. The end is to construct the heat kernel and the Green function of the sub-Laplacian, and the means to this end is to reduce this operator to a family of operators, known as twisted Laplacians, on $\mathbb{R}^2$ using the Fourier transform. Wigner transforms and Weyl transforms studied in Chapter 20 are the tools used in the constructions of the heat kernels and Green functions of the twisted Laplacians, which can then be transported back to the Heisenberg group in order to produce the heat kernel and Green function of the sub-Laplacian.

Mathematicians and physicists with names and contributions built into the text are highlighted in historical notes. Brevity notwithstanding, it is my hope that students can realize that mathematics is the culmination of efforts of many outstanding individuals over centuries. Exercises, intended to be done by students in order to better understand the material in the book, are appended at the end of every chapter.

Preface to the Second Edition

In STEM, *i.e.,* Science, Technology, Engineering, and Mathematics, books need to be revised on a regular basis in order to reflect new developments in the fields, while upholding the aims and scope of earlier editions. This is necessary for professionals to disseminate timely knowledge to future generations of young minds.

In this second edition of the book, a thorough attempt has been made to correct the mistakes, remove the redundancies, and clarify the text of the first edition. The bibliography and the index have been expanded. More exercises have been added to several chapters in this new edition. The bibliography in this new edition contains the updated sources in the first edition and a mild expansion of references that have shaped and will broaden my vision of partial differential equations. The amount of exercises is minimal because the temptation of loading a lot of drill exercises is resisted and my hope is that students can have the time and the energy to do all the exercises in the book.

Of particular relevance to this new edition is the addition of three new chapters, *i.e.*, Chapters 24–26 at the end of this edition. Chapter 24 is devoted to several topics in analysis such as the holomorphic continuation of the Riemann zeta-function to a meromorphic function on $\mathbb{C}$ with a simple pole at the real number 1. This is instrumental in the unfoldings of Chapters 25 and 26, which are devoted to the eigenvalue distribution of the heat semigroup and the inverse of a fourth-order partial differential operator dubbed the twisted bi-Laplacian. These three new chapters entail a substantial upgrading of Chapter 2 on the gamma function. To be specific, we need the holomorphic continuation of the gamma function from the positve real axis to a meromorphic function on $\mathbb{C}$ with simple poles at the origin and the negative integers.

Chapter 1

The Multi-Index Notation

We begin with the standard multi-index notation in the modern theory of partial differential equations. Let $x = (x_1, x_2, \ldots, x_n)$ and $y = (y_1, y_2, \ldots, y_n)$ be points in the Euclidean space $\mathbb{R}^n$. Then the dot product $x \cdot y$ of x and y is defined by

$$x \cdot y = \sum_{j=1}^{n} x_j y_j$$

and the norm $|x|$ of x is given by

$$|x| = \left(\sum_{j=1}^{n} x_j^2 \right)^{1/2}.$$

Let $\alpha = (\alpha_1, \alpha_2, \ldots, \alpha_n)$, where the entries $\alpha_1, \alpha_2, \ldots, \alpha_n$ are nonnegative integers. Then we call α a multi-index and we define its length $|\alpha|$ by

$$|\alpha| = \sum_{j=1}^{n} \alpha_j.$$

Let $x = (x_1, x_2, \ldots, x_n)$ be a point in $\mathbb{R}^n$ and let $\alpha = (\alpha_1, \alpha_2, \ldots, \alpha_n)$ be a multi-index. Then we define x^α by

$$x^\alpha = x_1^{\alpha_1} x_2^{\alpha_2} \cdots x_n^{\alpha_n}.$$

The simplest partial differential operators on $\mathbb{R}^n$ are obviously

$$\frac{\partial}{\partial x_1}, \frac{\partial}{\partial x_2}, \ldots, \frac{\partial}{\partial x_n}.$$

For $j = 1, 2, \ldots, n$, we let

$$\partial_j = \frac{\partial}{\partial x_j}$$

and

$$D_j = -i\partial_j,$$

DOI: 10.1201/9781003206781-1

1

where $i^2 = -1$. It will be seen later on in this book that the introduction of the factor $-i$ makes many formulas look much better. If $\alpha = (\alpha_1, \alpha_2, \ldots, \alpha_n)$ is a multi-index, then

$$\partial^\alpha = \partial_1^{\alpha_1} \partial_2^{\alpha_2} \cdots \partial_n^{\alpha_n}$$

and

$$D^\alpha = D_1^{\alpha_1} D_2^{\alpha_2} \cdots D_n^{\alpha_n}.$$

We denote by $C^\infty(\mathbb{R}^n)$ the set of all functions f on $\mathbb{R}^n$ such that $\partial^\alpha f$ is continuous on $\mathbb{R}^n$ for all multi-indices α. All functions in this book are assumed to be complex-valued unless noted otherwise.

The most general linear partial differential operator $P(x, D)$ of order m on $\mathbb{R}^n$ is of the form

$$P(x, D) = \sum_{|\alpha| \leq m} a_\alpha(x) D^\alpha,$$

where $a_\alpha \in C^\infty(\mathbb{R}^n)$. If

$$a_\alpha(x) = a_\alpha, \quad |\alpha| \leq m,$$

where a_α is a constant, then $P(x, D)$ is an operator with constant coefficients and we denote it by $P(D)$ and

$$P(D) = \sum_{|\alpha| \leq m} a_\alpha D^\alpha. \tag{1.1}$$

A simple observation is that there is an obvious one-to-one correspondence between the set of all linear partial differential operators

$$P(D) = \sum_{|\alpha| \leq m} a_\alpha D^\alpha$$

with constant coefficients and the corresponding set of polynomials P on $\mathbb{R}^n$ given by

$$P(\xi) = \sum_{|\alpha| \leq m} a_\alpha \xi^\alpha, \quad \xi \in \mathbb{R}^n.$$

The following formula of Leibniz gives a compact and explicit expression for the derivative of the product of two functions. It is reminiscent of the binomial theorem.

Proposition 1.1 *Let f and g be functions in $C^\infty(\mathbb{R}^n)$. Then for all multi-indices α,*

$$\partial^\alpha(fg) = \sum_{\beta \leq \alpha} \binom{\alpha}{\beta} (\partial^\beta f)(\partial^{\alpha-\beta} g),$$

where $\beta \leq \alpha$ means that $\beta_j \leq \alpha_j$, $j = 1, 2, \ldots, n$, and

$$\binom{\alpha}{\beta} = \binom{\alpha_1}{\beta_1} \binom{\alpha_2}{\beta_2} \cdots \binom{\alpha_n}{\beta_n}.$$

Proposition 1.1 is a special case of the following proposition, which can be found in [20, 36].

Proposition 1.2 *Let f and g be functions in $C^\infty(\mathbb{R}^n)$. Then for all partial differential operators $P(D)$ with constant coefficients given by (1.1),*

$$P(D)(fg) = \sum_{|\nu| \le m} \frac{1}{\mu!} (P^\mu(D)f)(D^\mu g),$$

where

$$\mu! = \mu_1! \mu_2! \cdots \mu_n!$$

and $P^\mu(D)$ is the operator with constant coefficients corresponding to the polynomial P^μ given by

$$P^\mu(\xi) = (\partial^\mu P)(\xi), \quad \xi \in \mathbb{R}^n.$$

For a proof of Proposition 1.2, we use the following lemma, which is a multidimensional version of Taylor's formula for polynomials.

Lemma 1.3 *Let P be a polynomial of degree m on $\mathbb{R}^n$. Then for all ξ and η in $\mathbb{R}^n$,*

$$P(\xi + \eta) = \sum_{|\mu| \le m} P^\mu(\xi) \eta^\mu.$$

Proof Using the one-dimensional Taylor's formula for polynomials on $\mathbb{R}$ repeatedly, we get

$$
\begin{aligned}
&P(\xi + \eta) \\
=\ & P(\xi_1 + \eta_1, \xi_2 + \eta_2, \xi_3 + \eta_3, \ldots, \xi_n + \eta_n) \\
=\ & \sum_{\mu_1} \frac{1}{\mu_1!} (\partial_1^{\mu_1} P)(\xi_1, \xi_2 + \eta_2, \xi_3 + \eta_3 \ldots, \xi_n) \eta_1^{\mu_1} \\
=\ & \sum_{\mu_1, \mu_2} \frac{1}{\mu_1! \mu_2!} (\partial_1^{\mu_1} \partial_2^{\mu_2} P)(\xi_1, \xi_2, \xi_3 + \eta_3, \ldots, \xi_n + \eta_n) \eta_1^{\mu_1} \eta_2^{\mu_2} \\
=\ & \cdots \\
=\ & \sum_{\mu_1, \mu_2, \ldots, \mu_n} \frac{1}{\mu_1! \mu_2! \cdots \mu_n!} (\partial_1^{\mu_1} \partial_2^{\mu_2} \cdots \partial_n^{\mu_n} P)(\xi_1, \xi_2, \ldots, \xi_n) \eta_1^{\mu_1} \eta_2^{\mu_2} \cdots \eta_n^{\mu_n} \\
=\ & \sum_{|\mu| \le m} \frac{1}{\mu!} (\partial^\mu P)(\xi) \eta^\mu.
\end{aligned}
$$

□

Proof of Proposition 1.2 We begin with the product formula for differentiation giving us

$$D_k(fg) = f D_k g + g D_k f, \quad k = 1, 2, \ldots, n.$$

Thus,

$$P(D)(fg) = \sum_{|\mu| \leq m} (Q_\mu(D)f)(D^\mu g), \tag{1.2}$$

where the partial differential operator $Q_\mu(D)$ is to be determined for $|\mu| \leq m$. Let f and g be functions on $\mathbb{R}^n$ given by

$$f(x) = e^{ix \cdot \xi}, \quad x \in \mathbb{R}^n,$$

and

$$g(x) = e^{ix \cdot \eta}, \quad x \in \mathbb{R}^n.$$

Then for $|\mu| \leq m$,

$$(D^\mu f)(x) = \xi^\mu f(x), \quad x \in \mathbb{R}^n, \tag{1.3}$$

and

$$(D^\mu g)(x) = \eta^\mu g(x), \quad x \in \mathbb{R}^n. \tag{1.4}$$

Moreover, for every polynomial Q on $\mathbb{R}^n$, we can use (1.3) and (1.4) to get

$$(Q(D)f)(x) = Q(\xi)f(x), \quad x \in \mathbb{R}^n, \tag{1.5}$$

and

$$(Q(D)g)(x) = Q(\eta)g(x), \quad x \in \mathbb{R}^n. \tag{1.6}$$

Since

$$(fg)(x) = e^{ix \cdot (\xi + \eta)} \quad x \in \mathbb{R}^n,$$

it follows from (1.5) and (1.6) that

$$P(D)(fg) = P(\xi + \eta)fg. \tag{1.7}$$

Applying (1.5) and (1.6) to (1.2), we get

$$P(D)(fg) = \left(\sum_{|\mu| \leq m} Q_\mu(\xi)\eta^\mu \right) fg. \tag{1.8}$$

So, by (1.7) and (1.8), we get

$$P(\xi + \eta) = \sum_{|\mu| \leq m} Q_\mu(\xi)\eta^\mu.$$

By Taylor's formula in Lemma 1.3, we obtain

$$Q_\mu(\xi) = \frac{1}{\mu!}(\partial^\mu P)(\xi)$$

and this completes the proof.   □

Historical Notes

Euclid (fl. 300 BC) was a Greek mathematician best remembered for his Euclidean geometry and the proof of the infinitude of prime numbers. His 13-volume treatise entitled "Elements" contains topics in geometry and number theory. The axiomatic methods in mathematics can be traced back to Euclid's Elements. The Euclidean space $\mathbb{R}^n$ has been and is still among the most important spaces on which modern analysis is built.

Gottfried Wilhelm Leibniz (1646–1716) was a German mathematician. He is usually considered as an inventor of differential and integral calculus independently of Isaac Newton. His notation for the derivatives and integrals, among others, has contributed much to the dissemination of the subject.

Brooke Taylor (1685–1731), an English mathematician, is best known for the Taylor series and the Taylor theorem to students in analysis.

Exercises

1. Prove that for all x in $\mathbb{R}^n$ and all multi-indices α,

$$|x^\alpha| \le |x|^{|\alpha|}.$$

2. Let α and β be multi-indices such that $\beta \le \alpha$. Prove that

$$\partial^\beta x^\alpha = \binom{\alpha}{\beta}\beta! x^{\alpha-\beta}. \tag{1.9}$$

   What happens if β is not $\le \alpha$?

3. For all nonnegative integers N and all x in $\mathbb{R}^n$, prove that

$$|x|^{2N} \le n^N \sum_{|\alpha|=N} |x^\alpha|^2.$$

4. Let l be a fixed positive integer. How many multi-indices are there in $\mathbb{R}^n$ with length l?

5. Prove Proposition 1.1.

Chapter 2

The Gamma Function

One of the the most important special functions in mathematics is undoubtedly the gamma function of Euler. It is the function Γ on $(0, \infty)$ defined by

$$\Gamma(x) = \int_0^\infty e^{-t} t^{x-1} dt, \quad x > 0.$$

Example 2.1 Compute $\Gamma(1)$.

Solution By definition,

$$\Gamma(1) = \int_0^\infty e^{-t} dt = -e^{-t}|_0^\infty = 1.$$

$\square$

Example 2.2 Compute $\Gamma(\frac{1}{2})$.

Solution By definition,

$$\Gamma\left(\frac{1}{2}\right) = \int_0^\infty e^{-t} t^{-1/2} dt.$$

If we let $t = x^2$, then $x = \sqrt{t}$ and $dx = \frac{1}{2} t^{-1/2} dt$. Hence

$$\Gamma\left(\frac{1}{2}\right) = 2 \int_0^\infty e^{-x^2} dx = 2\frac{\sqrt{\pi}}{2} = \sqrt{\pi}.$$

$\square$

In order to compute $\Gamma(x)$ for other values of x, we use the recurrence formula in the following theorem.

Theorem 2.3 $\Gamma(x + 1) = x\Gamma(x), \quad x > 0.$

Proof Using the definition of the gamma function, we get

$$\Gamma(x + 1) = \int_0^\infty e^{-t} t^x dt = -e^{-t} t^x|_0^\infty + x \int_0^\infty e^{-t} t^{x-1} dt = x\Gamma(x)$$

DOI: 10.1201/9781003206781-2

for all positive real numbers x. $\qquad\qquad\qquad\qquad\qquad\qquad$ □

We can use Theorem 2.3 to extend the definition of the gamma function from positive values of x to all values of x for which $x \neq -n$, $n = 0, 1, 2, \ldots$. To wit, let $x \in (-1, 0)$. Then $x + 1 \in (0, 1)$. Therefore, $\Gamma(x + 1)$ is defined and we can define $\Gamma(x)$ by

$$\Gamma(x) = \frac{\Gamma(x+1)}{x}.$$

If $x \in (-2, -1)$, then $x + 1 \in (-1, 0)$. Since $\Gamma(x + 1)$ is defined, it follows that we can define $\Gamma(x)$ by

$$\Gamma(x) = \frac{\Gamma(x+1)}{x}.$$

Repeating this argument, we can define $\Gamma(x)$ for all x such that $x \neq -n$, $n = 0, 1, 2, \ldots$.

Example 2.4 Compute $\Gamma(-\frac{3}{2})$.

Solution Using the extension of the gamma function to negative values, we get

$$
\begin{aligned}
\Gamma\left(-\frac{3}{2}\right) &= \frac{\Gamma\left(-\frac{3}{2}+1\right)}{-\frac{3}{2}} = \frac{\Gamma\left(-\frac{1}{2}\right)}{-\frac{3}{2}} \\
&= \frac{\Gamma\left(-\frac{1}{2}+1\right)}{\left(-\frac{3}{2}\right)\left(-\frac{1}{2}\right)} = \frac{4}{3}\Gamma\left(\frac{1}{2}\right) = \frac{4\sqrt{\pi}}{3}.
\end{aligned}
$$

$\qquad\qquad\qquad\qquad\qquad\qquad$ □

The following theorem brings out the significance of the gamma function.

Theorem 2.5 *Let n be a positive integer. Then $\Gamma(n + 1) = n!$.*

Proof Using the recurrence formula in Theorem 2.3, we get

$$
\begin{aligned}
\Gamma(n+1) &= n\Gamma(n) = n(n-1)\Gamma(n-1) \\
&= n(n-1)(n-2)\Gamma(n-2) \\
&= n(n-1)(n-2)\cdots 2 \cdot 1\Gamma(1).
\end{aligned}
$$

Since $\Gamma(1) = 1$, it follows that $\Gamma(n + 1) = n!$. $\qquad\qquad\qquad$ □

Theorem 2.5 suggests that the gamma function extends the factorial function from positive integers to all values of x for which $x \neq -n$, $n = 1, 2, \ldots$. To see this better, we make the following definition.

Definition 2.6 For $x \neq -n, n = 1, 2, \ldots$, we define $x!$ by

$$x! = \Gamma(x + 1).$$

Using Definition 2.6, we see that there is deeper truth than just mere convention that $0! = 1$. Indeed, by Definition 2.6,

$$0! = \Gamma(0 + 1) = \Gamma(1) = 1.$$

Example 2.7 Compute $\frac{1}{2}!$.

Proof By Definition 2.6, we get

$$\frac{1}{2}! = \Gamma\left(\frac{1}{2} + 1\right) = \frac{1}{2}\Gamma\left(\frac{1}{2}\right) = \frac{\sqrt{\pi}}{2}.$$

$\square$

As a nontrivial application of the gamma function, let us compute the surface area $|\mathbb{S}^{n-1}|$ of the unit sphere $\mathbb{S}^{n-1}$ with center at the origin in $\mathbb{R}^n$. Before we do this, let us note that in polar coordinates in $\mathbb{R}^n$,

$$dx = r^{n-1} dr\, d\sigma,$$

where $r = |x| = \sqrt{\sum_{j=1}^n x_j^2}$ and $d\sigma$ is the surface measure on $\mathbb{S}^{n-1}$. Since

$$dx = r dr\, d\theta, \quad r \in (0, \infty), \theta \in [0, 2\pi),$$

in $\mathbb{R}^2$ and

$$dx = r^2 \sin\phi\, dr\, d\phi\, d\theta, \quad r \in (0, \infty), \phi \in [0, \pi], \theta \in [0, 2\pi),$$

in $\mathbb{R}^3$, it follows that

$$d\sigma = d\theta, \quad \theta \in [0, 2\pi),$$

in $\mathbb{R}^2$ and

$$d\sigma = \sin\phi\, d\phi\, d\theta, \quad \phi \in [0, \pi], \theta \in [0, 2\pi),$$

in $\mathbb{R}^3$. The polar coordinates in $\mathbb{R}^3$ are better known as spherical coordinates.

Theorem 2.8 $|\mathbb{S}^{n-1}| = \frac{2\pi^{n/2}}{\Gamma(\frac{n}{2})}$.

Proof Since

$$\int_{\mathbb{R}^n} e^{-|x|^2} dx = \pi^{n/2},$$

we can use polar coordinates in $\mathbb{R}^n$ to get

$$\pi^{n/2} = \int_{\mathbb{R}^n} e^{-|x|^2} dx = \int_0^\infty \int_{\mathbb{S}^{n-1}} e^{-r^2} r^{n-1} dr\, d\sigma = |\mathbb{S}^{n-1}| \int_0^\infty e^{-r^2} r^{n-1} dr.$$

If we let $t = r^2$, then $r = \sqrt{t}$, $dr = \frac{1}{2} t^{-1/2} dt$, and

$$\int_0^\infty e^{-r^2} r^{n-1} dr = \frac{1}{2} \int_0^\infty e^{-t} t^{(n/2)-1} dt = \frac{1}{2}\Gamma\left(\frac{n}{2}\right).$$

Therefore

$$|\mathbb{S}^{n-1}| = \frac{2\pi^{n/2}}{\Gamma\left(\frac{n}{2}\right)},$$

as asserted. ☐

That Γ is a holomorphic function on the open right half plane $\{s \in \mathbb{C} : \operatorname{Re} s > 0\}$ is easy to show and is left as an exercise. Mimicking the same technique for continuing the definition of the real-valued gamma function, initially defined on $(0, \infty)$ to a continuous function on $\mathbb{R} \setminus \{-n : n = 0, 1, 2, \dots\}$, we have the following theorem.

Theorem 2.9 *The holomorphic function Γ, initially defined on the open right half plane $\{s \in \mathbb{C} : \operatorname{Re} s > 0\}$, can be holomorphically continued to a meromorphic function on $\mathbb{C} \setminus \{-n : n = 0, 1, 2, \dots\}$ with a simple pole only at $-n$, $n = 0, 1, 2, \dots$.*

We need some holomorphic versions of the identities for the gamma function. The following lemma is needed.

Lemma 2.10 *Let $a \in (0, 1)$. Then*

$$\int_0^\infty \frac{x^{a-1}}{1+x} dx = \frac{\pi}{\sin(\pi a)}.$$

The proof of Lemma 2.10 is a basic exercise in complex analysis and is best left as an exercise. We can now prove the following identity for the gamma function.

Theorem 2.11 *For each s in $\mathbb{C}$ that is not an integer,*

$$\Gamma(s)\Gamma(1-s) = \frac{\pi}{\sin(\pi s)}.$$

Proof Let $s \in (0, 1)$. Then

$$\Gamma(1-s) = \int_0^\infty e^{-u} u^{-s} du.$$

Let t be a positive number. Then we change the variable of integration from u to v via $vt = u$. Hence

$$\Gamma(1-s) = t \int_0^\infty e^{-vt}(vt)^{-s} dv.$$

Therefore

$$\begin{aligned}
\Gamma(1-s)\Gamma(s) &= \int_0^\infty e^{-t} t^s \left(t \int_0^\infty e^{-vt}(vt)^{-s} dv \right) \frac{dt}{t} \\
&= \int_0^\infty \int_0^\infty e^{-t(1+v)} v^{-s} dv \, dt. \\
&= \int_0^\infty \left(\int_0^\infty e^{-t(1+v)} dt \right) dv.
\end{aligned}$$

But

$$\int_0^\infty e^{-t(1+v)}dt = \frac{e^{-t(1+v)}}{1+v}\bigg|_0^\infty = \frac{1}{1+v}.$$

Therefore by Lemma 2.10,

$$\Gamma(s)\Gamma(1-s) = \int_0^\infty \frac{v^{-s}}{1+v}dv = \frac{\pi}{\sin(\pi(1-s))} = \frac{\pi}{\sin(\pi s)}.$$

By Theorem 2.9, $\Gamma(1-s)\Gamma(s)$ is a meromorphic function on $\mathbb{C}$ with a simple pole at each integer. The same is true for the function $\frac{\pi}{\sin(\pi s)}$. Therefore by holomorphic continuation,

$$\Gamma(1-s)\Gamma(s) = \frac{\pi}{\sin(\pi s)}, \quad s \in \mathbb{C} \setminus \{0, \pm 1, \pm 2, \dots\}.$$

$\square$

The following Legendre duplication formula for the gamma function is to be used in the last three chapters of the book.

Theorem 2.12 *Let s be a complex number such that*

$$s \notin \{-n/2 : n = 0, 1, 2, \dots\}.$$

Then

$$2^{1-2s}\sqrt{\pi}\,\Gamma(2s) = \Gamma(s)\Gamma\left(s + \frac{1}{2}\right).$$

Proof Let B be the function defined by

$$B(z, w) = \int_0^1 t^{z-1}(1-t)^{w-1}dt$$

for all z and w in $\mathbb{C}$ with $\operatorname{Re} z > 0$ and $\operatorname{Re} w > 0$. Then

$$B(z, w) = \frac{\Gamma(z)\Gamma(w)}{\Gamma(z+w)}. \tag{2.1}$$

Indeed, let x and y be positive numbers. Then

$$
\begin{aligned}
\Gamma(x)\Gamma(y) &= \left(\int_0^\infty e^{-u}u^{x-1}du\right)\left(\int_0^\infty e^{-v}v^{y-1}dv\right) \\
&= \int_0^\infty \int_0^\infty e^{-(u+v)}u^{x-1}v^{y-1}du\,dv.
\end{aligned}
$$

Let $u = \delta t$ and $v = \delta(1-t)$. Since

$$\left|\det\begin{pmatrix} \frac{\partial u}{\partial t} & \frac{\partial u}{\partial \delta} \\ \frac{\partial v}{\partial t} & \frac{\partial v}{\partial \delta} \end{pmatrix}\right| = \left|\det\begin{pmatrix} \delta & t \\ -\delta & 1-t \end{pmatrix}\right| = \delta,$$

we get

$$du \, dv = \delta \, dt \, d\delta.$$

Therefore

$$
\begin{aligned}
\Gamma(x)\Gamma(y) &= \int_0^\infty \int_0^1 e^{-\delta}(\delta t)^{x-1}(\delta(1-t))^{y-1}\delta \, dt \, d\delta \\
&= \left(\int_0^\infty e^{-\delta}\delta^{x+y-1}d\delta \right) \left(\int_0^1 t^{x-1}(1-t)^{y-1}dt \right) \\
&= \Gamma(x+y)B(x,y).
\end{aligned}
$$

Therefore (2.1) is established if we invoke Theorem 2.9. Now, we let $x = y = s$ in (2.1). Then

$$\frac{\Gamma(s)\Gamma(s)}{\Gamma(2s)} = \int_0^1 u^{s-1}(1-u)^{s-1}du.$$

Let $u = \sin^2 \theta$. Then $du = 2\sin\theta\cos\theta \, d\theta$. So,

$$
\begin{aligned}
\frac{\Gamma(s)\Gamma(s)}{\Gamma(2s)} &= \int_0^{\pi/2} \sin^{2s-2}\theta \, (1 - \sin^2\theta)^{s-1} 2\sin\theta\cos\theta \, d\theta \\
&= \int_0^{\pi/2} 2\sin^{2s-1}\theta \cos^{2s-1}\theta \, d\theta \\
&= 2^{2(1-s)} \int_0^{\pi/2} 2^{2s-1} \sin^{2s-1}\theta \cos^{2s-1}\theta \, d\theta \\
&= 2^{2(1-s)} \int_0^{\pi/2} \sin^{2s-1}(2\theta) \, d\theta.
\end{aligned}
$$

Now, let $\phi = 2\theta$. Then

$$
\begin{aligned}
& \frac{\Gamma(s)\Gamma(s)}{\Gamma(2s)} \\
&= 2^{1-2s} \int_0^{\pi} \sin^{2s-1}\phi \, d\phi \\
&= 2^{1-2s} \int_0^{\pi/2} \sin^{2s-1}\phi \, d\phi + 2^{1-2s} \int_{\pi/2}^{\pi} \sin^{2s-1}\phi \, d\phi. \qquad (2.2)
\end{aligned}
$$

For the last integral in the preceding line, let $\omega = \phi - \frac{\pi}{2}$. Then

$$\frac{\Gamma(s)\Gamma(s)}{\Gamma(2s)} = 2^{1-2s} \int_0^{\pi/2} \sin^{2s-1}\phi \, d\phi + 2^{1-2s} \int_0^{\pi/2} \cos^{2s-1}\omega \, d\omega.$$

We leave it as an exercise to prove that

$$B(z,w) = \int_0^{\pi/2} 2\sin^{2z-1}\theta \cos^{2w-1}\theta \, d\theta \qquad (2.3)$$

for all complex numbers z and w with $\operatorname{Re} z > 0$ and $\operatorname{Re} w > 0$. Thus, by (2.2) and (2.3),

$$
\begin{aligned}
\frac{\Gamma(s)\Gamma(s)}{\Gamma(2s)} &= 2^{-2s} \int_0^{\pi/2} 2 \sin^{2s-1}\phi \, \cos^{[2(1/2)]-1}\phi \, d\phi \\
&\quad + 2^{-2s} \int_0^{\pi/2} 2 \sin^{[2(1/2)]-1}\omega \, \cos^{2s-1}\omega \, d\omega \\
&= 2^{-2s}(B(s,1/2) + B(1/2,s)) = 2^{1-2s} B(s,1/2)
\end{aligned}
$$

if we note that $B(z,w) = B(w,z)$ for all complex numbers z and w with $\operatorname{Re} z > 0$ and $\operatorname{Re} w > 0$. Therefore

$$
\frac{\Gamma(s)\Gamma(s)}{\Gamma(2s)} = 2^{1-2s}\frac{\Gamma(s)\Gamma\left(\frac{1}{2}\right)}{\Gamma\left(s+\frac{1}{2}\right)}.
$$

Since $\Gamma\left(\frac{1}{2}\right) = \sqrt{\pi}$, it follows that

$$
2^{1-2s}\sqrt{\pi}\,\Gamma(2s) = \Gamma(s)\Gamma\left(s+\frac{1}{2}\right) \tag{2.4}
$$

for all s in $\{s \in \mathbb{C} : \operatorname{Re} s > 0\}$. Since both sides of (2.4) are holomorphic functions on $\mathbb{C}\backslash\{-n/2 : n = 0, 1, 2, \dots\}$, the proof is complete by holomorphic continuation.

$\square$

Remark 2.13 The 39-page book [3] is a beautiful account of the gamma function.

Historical Notes

Leonhard Euler (1707–1783) is usually considered the greatest mathematician of the 18th century. He was brought up near Basel in Switzerland. He was a prolific mathematician. In addition to his many contributions to analysis such as the gamma function, we mention his mathematical formulation of problems in mechanics and his techniques of solving these mathematical problems.

Adrien-Marie Legendre (1752–1833) was a French mathematician. Besides the very useful duplication formula for the gamma function named after him, his contributions to analysis include elliptic integrals and Legendre polynomials in spherical harmonics.

Exercises

1. Compute $\frac{5}{2}!$ and $\left(-\frac{7}{2}\right)!$.

2. Prove that the graph of Γ is concave up on $(0, \infty)$.

3. Prove that Γ has a local minimum in the open interval $(1, 2)$.

4. Find $\lim_{x \to 0+} \Gamma(x)$ and $\lim_{x \to \infty} \Gamma(x)$.

5. Sketch the graph of Γ on $(0, \infty)$.

6. Is there a function $\tilde{\Gamma}$ defined on all of $\mathbb{R}$ such that $\tilde{\Gamma}(x) = \Gamma(x)$ and $\tilde{\Gamma}(x + 1) = x\tilde{\Gamma}(x)$ for all $x \neq -n$, $n = 0, 1, \ldots$?

7. Compute the volume of the unit ball in $\mathbb{R}^n$.

8. Prove that the gamma function Γ defined on the open right half plane $\{s \in \mathbb{C} : \operatorname{Re} s > 0\}$ is a holomorphic function.

9. Prove Theorem 2.9.

10. Prove Lemma 2.10.

11. Find all complex numbers s for which
$$\Gamma((1 - s)/2)\Gamma((1 + s)/2) = \frac{\pi}{\sin(\pi(1 - s)/2)}.$$

12. Find all complex zeros of the gamma function Γ.

13. Prove that for all complex numbers z and w with $\operatorname{Re} z > 0$ and $\operatorname{Re} w > 0$,
$$B(z, w) = \frac{\Gamma(z)\Gamma(w)}{\Gamma(z + w)}.$$

Chapter 3

Convolutions

The convolution comes up very often in formulas for solutions of partial differential equations. Let f and g be measurable functions on $\mathbb{R}^n$. Then the convolution $f * g$ of f and g is defined by

$$(f * g)(x) = \int_{\mathbb{R}^n} f(x - y)g(y) \, dy, \quad x \in \mathbb{R}^n,$$

provided that the integral exists. In order to know when the integral exists, it is convenient to introduce some standard classes of functions.

For $1 \le p < \infty$, we let $L^p(\mathbb{R}^n)$ be the set of all measurable functions f on $\mathbb{R}^n$ such that

$$\int_{\mathbb{R}^n} |f(x)|^p dx < \infty.$$

We take $L^\infty(\mathbb{R}^n)$ to be the set of all essentially bounded functions on $\mathbb{R}^n$. If $f \in L^p(\mathbb{R}^n)$, $1 \le p < \infty$, then we define the norm $\|f\|_p$ of f by

$$\|f\|_p = \left\{ \int_{\mathbb{R}^n} |f(x)|^p dx \right\}^{1/p}.$$

If $f \in L^\infty(\mathbb{R}^n)$, then we define the norm $\|f\|_\infty$ by

$$\|f\|_\infty = \inf\{M : m\{x \in \mathbb{R}^n : |f(x)| > M\} = 0\},$$

where $m\{\cdots\}$ denotes the Lebesgue measure of the set $\{\cdots\}$.

Remark 3.1 Of particular importance is the space $L^2(\mathbb{R}^n)$, which is a Hilbert space. This fact is important for us when we study partial differential equations in Chapters 14–23. For the sake of simplicity in notation, we denote the inner product in $L^2(\mathbb{R}^n)$ by $(\,,\,)$ for all positive integers n, and it is given by

$$(f, g) = \int_{\mathbb{R}^n} f(x)\overline{g(x)} \, dx$$

for all f and g in $L^2(\mathbb{R}^n)$. The space, which the inner product $(\,,\,)$ is referred to, is clear from the context.

We can now give a theorem on when the convolution exists.

DOI: 10.1201/9781003206781-3

Theorem 3.2 *If $f \in L^1(\mathbb{R}^n)$ and $g \in L^p(\mathbb{R}^n)$, $1 \le p \le \infty$, then $f * g$ exists and is a function in $L^p(\mathbb{R}^n)$. Furthermore,*

$$\|f * g\|_p \le \|f\|_1 \|g\|_p.$$

For $p = \infty$, the proof is easy. Indeed, by a simple change of variable, we get

$$
\begin{aligned}
\int_{\mathbb{R}^n} |f(x-y)g(y)| \, dy &\le \|g\|_\infty \int_{\mathbb{R}^n} |f(x-y)| \, dy \\
&= \|g\|_\infty \int_{\mathbb{R}^n} |f(y)| \, dy \\
&= \|f\|_1 \|g\|_\infty.
\end{aligned}
$$

For $1 \le p < \infty$, we give a proof based on Minkowski's inequality in integral form to the effect that if f is a measurable function on $\mathbb{R}^n \times \mathbb{R}^n$, then

$$\left\{ \int_{\mathbb{R}^n} \left| \int_{\mathbb{R}^n} f(x,y) \, dy \right|^p dx \right\}^{1/p} \le \int_{\mathbb{R}^n} \left\{ \int_{\mathbb{R}^n} |f(x,y)|^p dx \right\}^{1/p} dy.$$

Instead of giving a rigorous proof of Minkowski's inequality, we give a heuristic argument for why it is true. To do this, we note that $\int_{\mathbb{R}^n} f(\cdot, y) \, dy$ can be considered to be a sum of functions indexed by y. So, the left-hand side is in fact the L^p norm of a sum of functions, which is at most the sum of the L^p norms of the functions. But the sum of the L^p norms of the functions is just $\int_{\mathbb{R}^n} \left\{ \int_{\mathbb{R}^n} |f(x,y)|^p dx \right\}^{1/p} dy$. To prove the inequality in Theorem 3.2 for $1 \le p < \infty$, we note that by Minkowski's inequality in integral form,

$$
\begin{aligned}
\|f * g\|_p &= \left\{ \int_{\mathbb{R}^n} |(f * g)(x)|^p dx \right\}^{1/p} \\
&= \left\{ \int_{\mathbb{R}^n} \left| \int_{\mathbb{R}^n} f(x-y) \, g(y) \, dy \right|^p dx \right\}^{1/p} \\
&= \left\{ \int_{\mathbb{R}^n} \left| \int_{\mathbb{R}^n} f(y) \, g(x-y) \, dy \right|^p dx \right\}^{1/p} \\
&\le \int_{\mathbb{R}^n} \left\{ \int_{\mathbb{R}^n} |f(y)|^p |g(x-y)|^p dx \right\}^{1/p} dy \\
&= \int_{\mathbb{R}^n} |f(y)| \left\{ \int_{\mathbb{R}^n} |g(x-y)|^p dx \right\}^{1/p} dy \\
&= \int_{\mathbb{R}^n} |f(y)| \left\{ \int_{\mathbb{R}^n} |g(x)|^p dx \right\}^{1/p} dy \\
&= \|f\|_1 \|g\|_p.
\end{aligned}
$$

Remark 3.3 The inequality in Theorem 3.2 is known as Young's inequality.

That functions in $L^p(\mathbb{R}^n)$, $1 \le p < \infty$, can be approximated by convolutions can be seen from the following theorem.

Theorem 3.4 *Let φ be a function in $L^1(\mathbb{R}^n)$ such that*

$$\int_{\mathbb{R}^n} \varphi(x)\, dx = a.$$

For every positive number ε, let φ_ε be the function on $\mathbb{R}^n$ defined by

$$\varphi_\varepsilon(x) = \varepsilon^{-n} \varphi\left(\frac{x}{\varepsilon}\right), \quad x \in \mathbb{R}^n.$$

Then for every f in $L^p(\mathbb{R}^n)$, $1 \le p < \infty$,

$$\lim_{\varepsilon \to 0+} \|f * \varphi_\varepsilon - af\|_p = 0.$$

The family $\{\varphi_\varepsilon : \varepsilon > 0\}$ of functions is known as the Friedrich mollifier associated to the function φ.

The proof of Theorem 3.4 is based on the following result, which is known as the L^p continuity of translations.

Lemma 3.5 *Let $f \in L^p(\mathbb{R}^n)$, $1 \le p < \infty$. Then*

$$\lim_{x \to 0} \|f_x - f\|_p = 0,$$

where

$$f_x(y) = f(x + y), \quad y \in \mathbb{R}^n.$$

To prove Lemma 3.5, we let $C_0(\mathbb{R}^n)$ be the set of all continuous functions f on $\mathbb{R}^n$ such that the support $\text{supp}(f)$ of f is compact, where $\text{supp}(f)$ is the closure of the set on which f is not equal to zero. A basic result in real analysis is that $C_0(\mathbb{R}^n)$ is dense in $L^p(\mathbb{R}^n)$ for $1 \le p < \infty$.

Proof of Lemma 3.5 Let $f \in L^p(\mathbb{R}^n)$. Then for all positive numbers δ, there exists a function g in $C_0(\mathbb{R}^n)$ such that $\|f - g\|_p < \frac{\delta}{3}$. So, for all $x \in \mathbb{R}^n$,

$$\|f_x - f\|_p \le \|f_x - g_x\|_p + \|g_x - g\|_p + \|g - f\|_p < \frac{2\delta}{3} + \|g_x - g\|_p.$$

Now, suppose that $g(y) = 0$ for $|y| > R$. Then for $|x| \le 1$,

$$g(x + y) - g(y) = 0$$

whenever $|y| > R + 1$. So,

$$\|g_x - g\|_p^p = \int_{|y| \le R+1} |g(x+y) - g(y)|^p dy \to 0$$

as $x \to 0$. Hence, there exists a positive number η such that

$$|x| < \eta \Rightarrow \|g_x - g\|_p < \frac{\delta}{3}.$$

So,

$$|x| < \eta \Rightarrow \|f_x - f\|_p < \delta$$

and the proof is complete. $\qquad\square$

Proof of Theorem 3.4 The starting point is the observation that

$$\int_{\mathbb{R}^n} \varphi_\varepsilon(x)\, dx = \varepsilon^{-n} \int_{\mathbb{R}^n} \varphi\left(\frac{x}{\varepsilon}\right) dx = \int_{\mathbb{R}^n} \varphi(x)\, dx = a$$

for every positive number ε. Then for every x in $\mathbb{R}^n$,

$$
\begin{aligned}
(f * \varphi_\varepsilon)(x) - af(x) &= \int_{\mathbb{R}^n} (f(x-y) - f(x))\varphi_\varepsilon(y)\, dy \\
&= \varepsilon^{-n} \int_{\mathbb{R}^n} (f(x-y) - f(x))\varphi\left(\frac{y}{\varepsilon}\right) dy \\
&= \int_{\mathbb{R}^n} (f(x-\varepsilon y) - f(x))\varphi(y)\, dy.
\end{aligned}
$$

Then, by Minkowski's inequality in integral form, we get for all positive numbers ε

$$
\begin{aligned}
\|f * \varphi_\varepsilon - af\|_p &= \left\{ \int_{\mathbb{R}^n} \left| \int_{\mathbb{R}^n} (f(x-\varepsilon y) - f(x))\varphi(y)\, dy \right|^p dx \right\}^{1/p} \\
&\leq \int_{\mathbb{R}^n} \left\{ \int_{\mathbb{R}^n} |f(x-\varepsilon y) - f(x)|^p |\varphi(y)|^p dx \right\}^{1/p} dy \\
&= \int_{\mathbb{R}^n} |\varphi(y)| \left\{ \int_{\mathbb{R}^n} |f(x-\varepsilon y) - f(x)|^p dx \right\}^{1/p} dy \\
&= \int_{\mathbb{R}^n} |\varphi(y)| \, \|f_{-\varepsilon y} - f\|_p\, dy.
\end{aligned}
$$

Now, for each fixed y in $\mathbb{R}^n$, the L^p continuity of translations implies that

$$\|f_{-\varepsilon y} - f\|_p \to 0$$

as $\varepsilon \to 0$. Furthermore,

$$|\varphi(y)| \, \|f_{-\varepsilon y} - f\|_p \leq 2\|f\|_p |\varphi(y)|, \quad y \in \mathbb{R}^n.$$

Since $\varphi \in L^1(\mathbb{R}^n)$, the Lebesgue dominated convergence theorem implies that

$$\|f * \varphi_\varepsilon - af\|_p \to 0$$

as $\varepsilon \to 0$. $\qquad\square$

We can now use convolutions to study two extremely important function spaces. We denote by $C_0^\infty(\mathbb{R}^n)$ the set of all functions φ in $C^\infty(\mathbb{R}^n)$ such that $\text{supp}(\varphi)$ is compact. We let $\mathcal{S}$ be the set of all functions φ in $C^\infty(\mathbb{R}^n)$ such that for all multi-indices α and β,

$$\sup_{x \in \mathbb{R}^n} |x^\alpha (\partial^\beta \varphi)(x)| < \infty.$$

The space $\mathcal{S}$ is usually called the Schwartz space.

Remark 3.6 $C_0^\infty(\mathbb{R}^n) \neq \phi$. Indeed, let φ be the function on $\mathbb{R}^n$ defined by

$$\varphi(x) = \begin{cases} e^{-1/(1-|x|^2)}, & |x| < 1, \\ 0, & |x| \geq 1. \end{cases}$$

Then $\varphi \in C_0^\infty(\mathbb{R}^n)$. Obviously, $C_0^\infty(\mathbb{R}^n)$ is a subset of $\mathcal{S}$. Since the function φ on $\mathbb{R}^n$ defined by

$$\varphi(x) = e^{-|x|^2}, \quad x \in \mathbb{R}^n,$$

is in $\mathcal{S}$ but not in $C_0^\infty(\mathbb{R}^n)$, it follows that $C_0^\infty(\mathbb{R}^n)$ is a proper subset of $\mathcal{S}$. It is an easy exercise to check that $\mathcal{S} \subset L^p(\mathbb{R}^n)$, $1 \leq p \leq \infty$. Furthermore, for all multi-indices α and all functions φ in $\mathcal{S}$, $x^\alpha \varphi \in \mathcal{S}$ and $\partial^\alpha \varphi \in \mathcal{S}$.

There are very few explicit formulas for functions in $C_0^\infty(\mathbb{R}^n)$. The ones that are known are built on the standard example in Remark 3.6. Are there many functions in $C_0^\infty(\mathbb{R}^n)$? The following theorem tells us that there are surprisingly many functions in $C_0^\infty(\mathbb{R}^n)$.

Theorem 3.7 $C_0^\infty(\mathbb{R}^n)$ *is dense in* $L^p(\mathbb{R}^n)$, $1 \leq p < \infty$.

To prove Theorem 3.7, we need two lemmas.

Lemma 3.8 *Let f and g be functions in $C_0(\mathbb{R}^n)$. Then $\text{supp}(f*g)$ is compact.*

Proof Let $x \in \mathbb{R}^n$ be such that $(f * g)(x) \neq 0$. Then

$$(f * g)(x) = \int_{\mathbb{R}^n} f(x - y)\, g(y)\, dy = \int_{\text{supp}(g)} f(x - y)\, g(y)\, dy \neq 0.$$

This implies that there exists a point y in $\text{supp}(g)$ such that $x - y \in \text{supp}(f)$. So,

$$x = (x - y) + y \in \text{supp}(f) + \text{supp}(g).$$

This gives us

$$\text{supp}(f * g) \subseteq \text{supp}(f) + \text{supp}(g)$$

and the proof is complete. □

Lemma 3.9 *Let $f \in L^p(\mathbb{R}^n)$, $1 \leq p \leq \infty$, and let $\varphi \in \mathcal{S}$. Then $f * \varphi \in C^\infty(\mathbb{R}^n)$ and for all multi-indices α,*

$$\partial^\alpha(f * \varphi) = f * (\partial^\alpha \varphi).$$

Proof We note that

$$(f * \varphi)(x) = \int_{\mathbb{R}^n} f(x - y)\varphi(y)\, dy = \int_{\mathbb{R}^n} f(y)\varphi(x - y)\, dy, \quad x \in \mathbb{R}^n.$$

So,

$$(\partial^\alpha(f * \varphi))(x) = \int_{\mathbb{R}^n} f(y)(\partial^\alpha \varphi)(x - y)\, dy, \quad x \in \mathbb{R}^n,$$

provided that the order of integration and differentiation can be interchanged. But, by Hölder's inequality,

$$\int_{\mathbb{R}^n} |f(y)|\,|(\partial^\alpha \varphi)(x - y)|\, dy \leq \|f\|_p \|\partial^\alpha \varphi\|_{p'},$$

where p' is the conjugate index of p. Hence, the interchange of the order of integration and differentiation is justified. □

Proof of Theorem 3.7 Let $g \in C_0(\mathbb{R}^n)$. Let $\varphi \in C_0^\infty(\mathbb{R}^n)$ be such that $\int_{\mathbb{R}^n} \varphi(x)\, dx = 1$. Then, by Theorem 3.4, $g * \varphi_\varepsilon \to g$ in $L^p(\mathbb{R}^n)$ as $\varepsilon \to 0$. Let $f \in L^p(\mathbb{R}^n)$. Then for every positive number δ, we can choose a function g in $C_0(\mathbb{R}^n)$ such that

$$\|f - g\|_p < \frac{\delta}{2}.$$

Let ε be a positive number such that

$$\|g * \varphi_\varepsilon - g\|_p < \frac{\delta}{2}.$$

Then

$$\|g * \varphi_\varepsilon - f\|_p \leq \|g * \varphi_\varepsilon - g\|_p + \|g - f\|_p < \frac{\delta}{2} + \frac{\delta}{2} = \delta.$$

By Lemmas 3.8 and 3.9, $g * \varphi_\varepsilon \in C_0^\infty(\mathbb{R}^n)$ and hence the proof is complete. □

Corollary 3.10 $\mathcal{S}$ *is dense in $L^p(\mathbb{R}^n)$, $1 \leq p < \infty$.*

Remark 3.11 The L^p continuity of translations, the approximations of functions in $L^p(\mathbb{R}^n)$ by means of the Friedrich mollifiers, and the density of $C_0^\infty(\mathbb{R}^n)$ and hence $\mathcal{S}$ in $L^p(\mathbb{R}^n)$ all depend on the basic fact that $C_0(\mathbb{R}^n)$ is dense in $L^p(\mathbb{R}^n)$ for $1 \leq p < \infty$. That $C_0(\mathbb{R}^n)$ is not dense in $L^\infty(\mathbb{R}^n)$ can be seen by looking at the function f on $\mathbb{R}^n$ given by

$$f(x) = 1, \quad x \in \mathbb{R}^n.$$

Indeed, suppose that for every positive number ε, we can find a function φ in $C_0(\mathbb{R}^n)$ such that

$$\|f - \varphi\|_\infty < \varepsilon.$$

Then

$$|\varphi(x) - 1| < \varepsilon, \quad x \in \mathbb{R}^n.$$

So, if we let $x \in \mathbb{R}^n \setminus \text{supp}(\varphi)$, we get $1 < \varepsilon$ for every positive number ε. This is a contradiction. Thus, $C_0^\infty(\mathbb{R}^n)$ and $\mathcal{S}$ cannot be dense in $L^\infty(\mathbb{R}^n)$. Is there an approximation theorem for functions in $L^\infty(\mathbb{R}^n)$ in terms of the Friedrich mollifiers? The following theorem gives some information in this direction.

Theorem 3.12 *Let φ and φ_ε be as in Theorem 3.4. Let f be a bounded function on $\mathbb{R}^n$ which is continuous on an open subset V of $\mathbb{R}^n$. Then*

$$f * \varphi_\varepsilon \to af$$

uniformly on every compact subset of V as $\varepsilon \to 0$.

Proof Let K be a compact subset of V. Then we need to prove that

$$\sup_{x \in K} |(f * \varphi_\varepsilon)(x) - af(x)| \to 0$$

as $\varepsilon \to 0$. But for all positive numbers ε and R, and all x in $\mathbb{R}^n$,

$$
\begin{aligned}
|(f * \varphi_\varepsilon)(x) - af(x)| &\leq \int_{\mathbb{R}^n} |f(x - \varepsilon y) - f(x)| \, |\varphi(y)| \, dy \\
&= \left(\int_{|y| \leq R} + \int_{|y| \geq R} \right) |f(x - \varepsilon y) - f(x)| \, |\varphi(y)| \, dy.
\end{aligned}
$$

Let $M = \sup_{x \in \mathbb{R}^n} |f(x)|$. Then for every positive number δ, we can find a positive number R_0 such that

$$\int_{|y| \geq R_0} |f(x - \varepsilon y) - f(x)| \, |\varphi(y)| \, dy \leq 2M \int_{|y| \geq R_0} |\varphi(y)| \, dy < \frac{\delta}{2}$$

for all positive numbers ε and all x in $\mathbb{R}^n$. Now, let ε_1 be a positive number such that the compact set K_{ε_1, R_0} given by

$$K_{\varepsilon_1, R_0} = K + \{z \in \mathbb{R}^n : |z| \leq \varepsilon_1 R_0\}$$

is contained in V. Since f is continuous on V and hence uniformly continuous on K_{ε_1, R_0}, it follows that there exists a positive number ε_2 such that

$$\int_{|y| \leq R_0} |f(z) - f(w)| \, |\varphi(y)| \, dy < \frac{\delta}{2}$$

for all z and w in K_{ε_1, R_0} with $|z - w| < \varepsilon_2$. Let $\varepsilon_0 = \min\left(\varepsilon_1, \frac{\varepsilon_2}{R_0}\right)$. If $\varepsilon < \varepsilon_0$, $x \in K$, and $|y| \leq R_0$, then

$$\int_{|y| \leq R_0} |f(x - \varepsilon y) - f(x)| \, |\varphi(y)| \, dy < \frac{\delta}{2}.$$

Thus,

$$\varepsilon < \varepsilon_0 \Rightarrow \sup_{x \in K} |(f * \varphi_\varepsilon)(x) - a f(x)| < \delta.$$

This completes the proof. $\square$

Historical Notes

Henri Léon Lebesgue (1875–1941) was a French mathematician. His theory of integration has become one of the core topics for every student in mathematics. This theory was originally published in his dissertation titled "Intégrale, longueur, aire" at the University of Nancy in 1902.

William Henry Young (1863–1942) was an English mathematician. He was well known for his works in analysis. In particular, Young's inequality is usually referred to as either the one in this book or the one on the product of two nonnegative numbers. More precisely, if a and b are two nonnegative numbers and p and p' are two positive numbers such that

$$\frac{1}{p} + \frac{1}{p'} = 1,$$

then

$$ab \leq \frac{a^p}{p} + \frac{b^{p'}}{p'},$$

and equality holds if and only if $a^p = b^{p'}$.

Hermann Minkowski (1864–1909) was a German mathematician. Despite the usefulness of Minkowski's inequality, his best work is certainly the Minkowski four-dimensional space-time, which provides the correct underpinnings of the special theory of relativity.

Laurent Schwartz (1915–2002) was a French mathematician. His most important work is on distributions that provide the mathematical foundation for the modern study in partial differential equations. He was awarded the Fields Medal in 1950 for his distribution theory. The Fields Medal is universally considered the equivalent of the Nobel Prize. There is no Nobel Prize for Mathematics.

Otto Hölder (1859–1967) was a German mathematician who obtained his doctoral degree from the University of Tübingen in 1882. In addition to the widely used Hölder's inequality, his works in analysis include the gamma function, potential theory, and the Hölder's condition on smoothness of functions.

Kurt Otto Friedrich (1901–1982) was born in Germany and passed away in the United States. He left Germany for the United States in 1937 and was a major figure in establishing the prestigious Courant Institute of Mathematical Sciences at New York University. His main contributions in mathematics are in partial differential equations.

Exercises

1. Let $f \in L^1(\mathbb{R}^n)$ and let g be the function on $\mathbb{R}^n$ given by

$$g(x) = 1, \quad x \in \mathbb{R}^n.$$

   Compute $f * g$.

2. Let f and g be functions in $L^1(\mathbb{R}^n)$. Prove that

$$\int_{\mathbb{R}^n} (f * g)(x)\, dx = \left(\int_{\mathbb{R}^n} f(x)\, dx \right) \left(\int_{\mathbb{R}^n} g(x)\, dx \right).$$

3. Let $f \in L^p(\mathbb{R}^n)$ and $g \in L^{p'}(\mathbb{R}^n)$, where $1 \leq p \leq \infty$ and p' is the conjugate index of p, *i.e.*,

$$\frac{1}{p} + \frac{1}{p'} = 1.$$

   (a) Prove that the function $f * g$ on $\mathbb{R}^n$ defined by

$$(f * g)(x) = \int_{\mathbb{R}^n} f(x - y)g(y)\, dy$$

   exists for all x in $\mathbb{R}^n$ and is a bounded function on $\mathbb{R}^n$.

   (b) Prove that $f * g$ is a continuous function on $\mathbb{R}^n$ if $1 \leq p < \infty$.

4. Let $\varepsilon > 0$. Construct a function φ_ε in $C_0^\infty(\mathbb{R}^n)$ such that

   (a) $0 \leq \varphi_\varepsilon(x) \leq 1, \quad x \in \mathbb{R}^n$,

   (b) $\varphi_\varepsilon(0) = 1$,

   (c) $\operatorname{supp}(\varphi_\varepsilon) = \{x \in \mathbb{R}^n : |x| \leq \varepsilon\}$.

5. Prove that for all multi-indices α and all functions φ in $\mathcal{S}$, $x^\alpha \varphi \in \mathcal{S}$ and $\partial^\alpha \varphi \in \mathcal{S}$.

6. Prove that if φ and ψ are functions in $\mathcal{S}$, then so is $\varphi * \psi$.

Chapter 4

Fourier Transforms

We give in this chapter a compact account of Fourier analysis that we need in this book. Fuller and more rigorous treatments can be found in the books [14, 45, 46, 58].

Let $f \in L^1(\mathbb{R}^n)$. Then we define the Fourier transform $\hat{f}$ of f to be the function on $\mathbb{R}^n$ by

$$\hat{f}(\xi) = (2\pi)^{-n/2} \int_{\mathbb{R}^n} e^{-ix\cdot\xi} f(x)\,dx, \quad \xi \in \mathbb{R}^n.$$

We sometimes denote $\hat{f}$ by $\mathcal{F}f$.

The first result tells us that the Fourier transform converts convolution into pointwise multiplication.

Proposition 4.1 *Let f and g be in $L^1(\mathbb{R}^n)$. Then*

$$(f * g)^\wedge = (2\pi)^{n/2} \hat{f}\hat{g}.$$

Proof By Theorem 3.2, $f * g \in L^1(\mathbb{R}^n)$. Then, using the definition of the Fourier transform, interchanging the order of integration, and changing the variable of integration, we get

$$
\begin{aligned}
(f * g)^\wedge(\xi) &= (2\pi)^{-n/2} \int_{\mathbb{R}^n} e^{-ix\cdot\xi}(f * g)(x)\,dx \\
&= (2\pi)^{-n/2} \int_{\mathbb{R}^n} e^{-ix\cdot\xi} \left(\int_{\mathbb{R}^n} f(x-y)g(y)\,dy \right) dx \\
&= (2\pi)^{-n/2} \int_{\mathbb{R}^n} e^{-iy\cdot\xi} g(y) \left(\int_{\mathbb{R}^n} e^{-i(x-y)\cdot\xi} f(x-y)\,dx \right) dy \\
&= (2\pi)^{-n/2} \int_{\mathbb{R}^n} e^{-iy\cdot\xi} g(y) \left(\int_{\mathbb{R}^n} e^{-ix\cdot\xi} f(x)\,dx \right) dy \\
&= (2\pi)^{n/2} \hat{f}(\xi)\hat{g}(\xi)
\end{aligned}
$$

for all ξ in $\mathbb{R}^n$. $\qquad\square$

Proposition 4.2 *Let $\varphi \in \mathcal{S}$. Then for every multi-index α, we have*

$$(D^\alpha \varphi)^\wedge(\xi) = \xi^\alpha \hat{\varphi}(\xi)$$

DOI: 10.1201/9781003206781-4

and

$$(D^\alpha \hat\varphi)(\xi) = ((-x)^\alpha \varphi)^\wedge(\xi)$$

for all ξ in $\mathbb{R}^n$. Moreover,

$$\hat\varphi \in \mathcal{S}.$$

Proof Integrating by parts, we get for all ξ in $\mathbb{R}^n$,

$$
\begin{aligned}
(D^\alpha \varphi)^\wedge(\xi) &= (2\pi)^{-n/2} \int_{\mathbb{R}^n} e^{-ix\cdot\xi}(D^\alpha \varphi)(x)\,dx \\
&= (2\pi)^{-n/2} \int_{\mathbb{R}^n} \xi^\alpha e^{-ix\cdot\xi}\varphi(x)\,dx \\
&= \xi^\alpha \hat\varphi(\xi).
\end{aligned}
$$

Interchanging the order of differentiation and integration, we get for all $\xi \in \mathbb{R}^n$,

$$
\begin{aligned}
(D^\alpha \hat\varphi)(\xi) &= (2\pi)^{-n/2} D^\alpha \int_{\mathbb{R}^n} e^{-ix\cdot\xi}\varphi(x)\,dx \\
&= (2\pi)^{-n/2} \int_{\mathbb{R}^n} (-x)^\alpha e^{-ix\cdot\xi}\varphi(x)\,dx \\
&= ((-x)^\alpha \varphi)^\wedge(\xi).
\end{aligned}
$$

Finally, let α and β be multi-indices. Then

$$
\begin{aligned}
\sup_{\xi\in\mathbb{R}^n} |\xi^\alpha (D^\beta \hat\varphi)(\xi)| &= \sup_{\xi\in\mathbb{R}^n} |\xi^\alpha ((-x)^\beta \varphi)^\wedge(\xi)| \\
&= \sup_{\xi\in\mathbb{R}^n} |(D^\alpha ((-x)^\beta \varphi))^\wedge(\xi)| \\
&\leq (2\pi)^{-n/2} \|D^\alpha ((-x)^\beta \varphi)\|_1 < \infty
\end{aligned}
$$

and the proof is complete. $\square$

While a function f in $L^1(\mathbb{R}^n)$ may be rather rough, the following result, which is usually called the Riemann–Lebesgue lemma, tells us that the Fourier transform of f is a continuous function vanishing at infinity on $\mathbb{R}^n$.

Proposition 4.3 *Let $f \in L^1(\mathbb{R}^n)$. Then $\hat f$ is continuous on $\mathbb{R}^n$ and*

$$\lim_{|\xi|\to\infty} \hat f(\xi) = 0.$$

We can use the following lemma to prove the Riemann–Lebesgue lemma.

Lemma 4.4 *Let $\{f_j\}_{j=1}^\infty$ be a sequence of functions in $L^1(\mathbb{R}^n)$ such that $f_j \to f$ in $L^1(\mathbb{R}^n)$ as $j \to \infty$. Then $\hat f_j \to \hat f$ uniformly on $\mathbb{R}^n$ as $j \to \infty$.*

Proof For all ξ in $\mathbb{R}^n$,

$$
\begin{aligned}
\sup_{\xi \in \mathbb{R}^n} |\widehat{f}_j(\xi) - \hat{f}(\xi)| &= (2\pi)^{-n/2} \left| \int_{\mathbb{R}^n} e^{-ix\cdot\xi} (f_j(x) - f(x)) \, dx \right| \\
&\leq (2\pi)^{-n/2} \int_{\mathbb{R}^n} |f_j(x) - f(x)| \, dx \\
&= (2\pi)^{-n/2} \|f_j - f\|_1 \to 0
\end{aligned}
$$

as $j \to \infty$. $\qquad\square$

Proof of Proposition 4.3 By the density of $\mathcal{S}$ in $L^1(\mathbb{R}^n)$, there exists a sequence $\{\varphi_j\}_{j=1}^{\infty}$ of functions in $\mathcal{S}$ such that $\varphi_j \to f$ in $L^1(\mathbb{R}^n)$ as $j \to \infty$. So, by Lemma 4.4, $\widehat{\varphi}_j \to \hat{f}$ uniformly on $\mathbb{R}^n$ as $j \to \infty$. Since $\widehat{\varphi}_j \in \mathcal{S}$ for $j = 1, 2, \ldots$, it follows that $\hat{f}$ is continuous on $\mathbb{R}^n$. To see that $\lim_{|\xi|\to\infty} |\hat{f}(\xi)| = 0$, let ε be a positive number. Then there exists a positive integer J such that

$$
|\widehat{\varphi}_J(\xi) - \hat{f}(\xi)| < \frac{\varepsilon}{2}, \quad \xi \in \mathbb{R}^n.
$$

Since $\lim_{|\xi|\to\infty} |\widehat{\varphi}_J(\xi)| = 0$, there exists a positive number R such that

$$
|\xi| > R \Rightarrow |\widehat{\varphi}_J(\xi)| < \frac{\varepsilon}{2}.
$$

Therefore

$$
|\xi| > R \Rightarrow |\hat{f}(\xi)| \leq |\widehat{\varphi}_J(\xi)| + |\hat{f}(\xi) - \widehat{\varphi}_J(\xi)| < \frac{\varepsilon}{2} + \frac{\varepsilon}{2} = \varepsilon.
$$

$\qquad\square$

We can also give a direct proof of Proposition 4.3.

Another Proof of Proposition 4.3 For continuity, we need to prove that for all ξ_0 in $\mathbb{R}^n$,

$$
\lim_{\xi \to \xi_0} \hat{f}(\xi) = \hat{f}(\xi_0).
$$

But

$$
\hat{f}(\xi) - \hat{f}(\xi_0) = (2\pi)^{-n/2} \int_{\mathbb{R}^n} (e^{-ix\cdot\xi} - e^{-ix\cdot\xi_0}) f(x) \, dx, \quad \xi \in \mathbb{R}^n.
$$

It is obvious that for all x in $\mathbb{R}^n$,

$$
e^{-ix\cdot\xi} - e^{-ix\cdot\xi_0} \to 0
$$

as $\xi \to \xi_0$ and

$$
|(e^{-ix\cdot\xi} - e^{-ix\cdot\xi_0}) f(x)| \leq 2|f(x)|, \quad x, \xi \in \mathbb{R}^n.
$$

Therefore by the Lebesgue dominated convergence theorem,

$$
\hat{f}(\xi) \to \hat{f}(\xi_0)
$$

as $\xi \to \xi_0$. For the limit, we note that if $\xi \neq 0$, then

$$
\begin{aligned}
\hat{f}(\xi) &= (2\pi)^{-n/2} \int_{\mathbb{R}^n} e^{-ix\cdot\xi} f(x)\, dx \\
&= -(2\pi)^{-n/2} \int_{\mathbb{R}^n} e^{-ix\cdot\xi + i\pi} f(x)\, dx \\
&= -(2\pi)^{-n/2} \int_{\mathbb{R}^n} e^{-i\xi\cdot\left(x - \frac{\pi\xi}{|\xi|^2}\right)} f(x)\, dx \\
&= -(2\pi)^{-n/2} \int_{\mathbb{R}^n} e^{-ix\cdot\xi} f\left(x + \frac{\pi\xi}{|\xi|^2}\right) dx.
\end{aligned}
$$

Thus, for $\xi \neq 0$,

$$
2\hat{f}(\xi) = (2\pi)^{-n/2} \int_{\mathbb{R}^n} e^{-ix\cdot\xi} \left\{ f(x) - f\left(x + \frac{\pi\xi}{|\xi|^2}\right) \right\} dx
$$

and hence for $\xi \neq 0$,

$$
2|\hat{f}(\xi)| \leq (2\pi)^{-n/2} \int_{\mathbb{R}^n} \left| f(x) - f\left(x + \frac{\pi\xi}{|\xi|^2}\right) \right| dx = (2\pi)^{-n/2} \|f - f_{\pi\xi/|\xi|^2}\|_1.
$$

So, by the L^1 continuity of translations in Lemma 3.5,

$$
\lim_{|\xi|\to\infty} \hat{f}(\xi) = 0.
$$

$\square$

The Euclidean structure of $\mathbb{R}^n$ allows us to define the translation, modulation, and dilation of functions defined on $\mathbb{R}^n$. To see how, let f be a function defined on $\mathbb{R}^n$. Then for all y in $\mathbb{R}^n$, we define the translation $T_y f$ and the modulation $M_y f$ of f with respect to y, respectively, by

$$
(T_y f)(x) = f(x + y), \quad x \in \mathbb{R}^n,
$$

and

$$
(M_y f)(x) = e^{ix\cdot y} f(x), \quad x \in \mathbb{R}^n.
$$

For every nonzero and real number a, we define the dilation $D_a f$ of f with respect to a by

$$
(D_a f)(x) = f(ax), \quad x \in \mathbb{R}^n.
$$

The interactions of Fourier transforms with translations, modulations, and dilations are formulated in the following proposition.

Proposition 4.5 *Let $f \in L^1(\mathbb{R}^n)$. Then for all y in $\mathbb{R}^n$ and all nonzero and real numbers a, we get*

$$
(T_y f)^\wedge = M_y \hat{f},
$$

$$
(M_y f)^\wedge = T_{-y} \hat{f},
$$

and

$$
(D_a f)^\wedge = |a|^{-n} D_{\frac{1}{a}} \hat{f}.
$$

Proof Obviously, $T_y f$, $M_y f$ and $D_a f$ are in $L^1(\mathbb{R}^n)$. By a simple change of variable, we get for all ξ in $\mathbb{R}^n$,

$$
\begin{aligned}
(T_y f)^\wedge(\xi) &= (2\pi)^{-n/2} \int_{\mathbb{R}^n} e^{-ix\cdot\xi}(T_y f)(x)\, dx \\
&= (2\pi)^{-n/2} \int_{\mathbb{R}^n} e^{-ix\cdot\xi} f(x+y)\, dx \\
&= (2\pi)^{-n/2} \int_{\mathbb{R}^n} e^{-i(x-y)\cdot\xi} f(x)\, dx \\
&= e^{iy\cdot\xi}(2\pi)^{-n/2} \int_{\mathbb{R}^n} e^{-ix\cdot\xi} f(x)\, dx \\
&= e^{iy\cdot\xi}\hat{f}(\xi) = (M_y \hat{f})(\xi).
\end{aligned}
$$

Also, for all ξ in $\mathbb{R}^n$,

$$
\begin{aligned}
(M_y f)^\wedge(\xi) &= (2\pi)^{-n/2} \int_{\mathbb{R}^n} e^{-ix\cdot\xi}(M_y f)(x)\, dx \\
&= (2\pi)^{-n/2} \int_{\mathbb{R}^n} e^{-ix\cdot\xi} e^{ix\cdot y} f(x)\, dx \\
&= \hat{f}(\xi - y) = (T_{-y}\hat{f})(\xi).
\end{aligned}
$$

Finally, by another change of variable, we get

$$
\begin{aligned}
(D_a f)^\wedge(\xi) &= (2\pi)^{-n/2} \int_{\mathbb{R}^n} e^{-ix\cdot\xi}(D_a f)(x)\, dx \\
&= (2\pi)^{-n/2} \int_{\mathbb{R}^n} e^{-ix\cdot\xi} f(ax)\, dx \\
&= (2\pi)^{-n/2} \int_{\mathbb{R}^n} e^{-i\left(\frac{x}{a}\right)\cdot\xi} f(x)|a|^{-n} dx \\
&= |a|^{-n}\hat{f}\left(\frac{\xi}{a}\right) = |a|^{-n}(D_{\frac{1}{a}}\hat{f})(\xi)
\end{aligned}
$$

for all ξ in $\mathbb{R}^n$. $\qquad\square$

Proposition 4.6 *Let φ be the function on $\mathbb{R}^n$ defined by*

$$\varphi(x) = e^{-|x|^2/2}, \quad x \in \mathbb{R}^n.$$

Then

$$\hat{\varphi}(\xi) = e^{-|\xi|^2/2}, \quad \xi \in \mathbb{R}^n.$$

Proof We first assume that $n = 1$. Then

$$\hat{\varphi}(\xi) = (2\pi)^{-1/2} \int_{-\infty}^{\infty} e^{-ix\xi}\varphi(x)\, dx = (2\pi)^{-1/2} \int_{-\infty}^{\infty} e^{-ix\xi}e^{-x^2/2} dx, \quad \xi \in \mathbb{R}.$$

Therefore, interchanging the order of differentiation and integration, and integrating by parts, we get

$$
\begin{aligned}
\hat{\varphi}'(\xi) &= -i(2\pi)^{-1/2} \int_{-\infty}^{\infty} e^{-ix\xi} x e^{-x^2/2} dx \\
&= i(2\pi)^{-1/2} \int_{-\infty}^{\infty} e^{-ix\xi} \left(\frac{d}{dx} e^{-x^2/2} \right) dx \\
&= -\xi(2\pi)^{-1/2} \int_{-\infty}^{\infty} e^{-ix\xi} e^{-x^2/2} dx \\
&= -\xi\hat{\varphi}(\xi)
\end{aligned}
$$

for all ξ in $\mathbb{R}$. Thus,

$$
\hat{\varphi}'(\xi) + \xi\hat{\varphi}(\xi) = 0, \quad \xi \in \mathbb{R},
$$

which is a first-order, linear ordinary differential equation in the unknown function $\hat{\varphi}$. An integrating factor is

$$
\mu(\xi) = e^{\int_0^\xi \eta \, d\eta} = e^{\xi^2/2}, \quad \xi \in \mathbb{R}.
$$

Thus,

$$
\frac{d}{d\xi} \left\{ e^{\xi^2/2} \hat{\varphi}(\xi) \right\} = 0, \quad \xi \in \mathbb{R},
$$

and we get

$$
e^{\xi^2/2} \hat{\varphi}(\xi) = C, \quad \xi \in \mathbb{R},
$$

where C is an arbitrary constant. To determine C, let $\xi = 0$ in the preceding formula and we get

$$
\begin{aligned}
C = \hat{\varphi}(0) &= (2\pi)^{-1/2} \int_{-\infty}^{\infty} e^{-x^2/2} dx \\
&= (2\pi)^{-1/2} \sqrt{2} \int_{-\infty}^{\infty} e^{-t^2} dt \\
&= \pi^{-1/2} \sqrt{\pi} = 1.
\end{aligned}
$$

Therefore

$$
\hat{\varphi}(\xi) = e^{-|\xi|^2/2}, \quad \xi \in \mathbb{R}.
$$

For $n > 1$, we get for all ξ in $\mathbb{R}^n$,

$$
\begin{aligned}
\hat{\varphi}(\xi) &= (2\pi)^{-n/2} \int_{\mathbb{R}^n} e^{-ix\cdot\xi} \varphi(x)\, dx \\
&= (2\pi)^{-n/2} \int_{\mathbb{R}^n} e^{-ix\cdot\xi} e^{-|x|^2/2} dx \\
&= (2\pi)^{-n/2} \int_{-\infty}^{\infty} \int_{-\infty}^{\infty} \cdots \int_{-\infty}^{\infty} e^{-i\sum_{j=1}^n x_j \xi_j} e^{-\frac{1}{2}\sum_{j=1}^n x_j^2} dx_1 dx_2 \cdots dx_n \\
&= \prod_{j=1}^n (2\pi)^{-1/2} \int_{-\infty}^{\infty} e^{-ix_j \xi_j} e^{-x_j^2/2} dx_j \\
&= \prod_{j=1}^n e^{-\xi_j^2/2} = e^{-|\xi|^2/2}.
\end{aligned}
$$

$\square$

We need one more property of the Fourier transform. It is known as the adjoint formula.

Proposition 4.7 *Let f and g be functions in $L^1(\mathbb{R}^n)$. Then*

$$
\int_{\mathbb{R}^n} \hat{f}(x) g(x)\, dx = \int_{\mathbb{R}^n} f(x) \hat{g}(x)\, dx.
$$

Proof Interchanging the order of integration, we get

$$
\begin{aligned}
\int_{\mathbb{R}^n} \hat{f}(x) g(x)\, dx &= (2\pi)^{-n/2} \int_{\mathbb{R}^n} \left(\int_{\mathbb{R}^n} e^{-ix\cdot y} f(y)\, dy \right) g(x)\, dx \\
&= (2\pi)^{-n/2} \int_{\mathbb{R}^n} f(y) \left(\int_{\mathbb{R}^n} e^{-ix\cdot y} g(x)\, dx \right) dy \\
&= \int_{\mathbb{R}^n} f(y) \hat{g}(y)\, dy.
\end{aligned}
$$

$\square$

We are now ready for an important result in Fourier analysis. First, we need a definition. For every function f in $L^1(\mathbb{R}^n)$, we define the inverse Fourier transform $\check{f}$ of f to be the function on $\mathbb{R}^n$ by

$$
\check{f}(x) = (2\pi)^{-n/2} \int_{\mathbb{R}^n} e^{ix\cdot\xi} f(\xi)\, d\xi, \quad x \in \mathbb{R}^n.
$$

The Fourier inversion formula is given in the following theorem.

Theorem 4.8 *Let $f \in \mathcal{S}$. Then $\hat{f}^{\vee} = f$.*

Proof We begin with the formula

$$(\hat{f}^{\vee})(x) = (2\pi)^{-n/2} \int_{\mathbb{R}^n} e^{ix\cdot\xi} \hat{f}(\xi)\, d\xi, \quad x \in \mathbb{R}^n. \tag{4.1}$$

For every positive number ε, we let I_ε be the function on $\mathbb{R}^n$ defined by

$$I_\varepsilon(x) = (2\pi)^{-n/2} \int_{\mathbb{R}^n} e^{ix\cdot\xi} e^{-\varepsilon^2|\xi|^2/2} \hat{f}(\xi)\, d\xi, \quad x \in \mathbb{R}^n. \tag{4.2}$$

Then, by Propositions 4.5 through 4.7, we get

$$I_\varepsilon(x) = (2\pi)^{-n/2}\varepsilon^{-n} \int_{\mathbb{R}^n} e^{-\frac{|x-\eta|^2}{2\varepsilon^2}} f(\eta)\, d\eta = (2\pi)^{-n/2}(f * \varphi_\varepsilon)(x), \quad x \in \mathbb{R}^n,$$

where $\varphi(x) = e^{-|x|^2/2}$, $x \in \mathbb{R}^n$. Since

$$\int_{\mathbb{R}^n} \varphi(x)\, dx = (2\pi)^{n/2},$$

it follows from Theorem 3.4 that for $1 \leq p < \infty$,

$$I_\varepsilon = (2\pi)^{-n/2}(f * \varphi_\varepsilon) \to f$$

in $L^p(\mathbb{R}^n)$ as $\varepsilon \to 0$. So, there exists a sequence $\{\varepsilon_j\}_{j=1}^\infty$ of positive numbers such that $I_{\varepsilon_j} \to f$ a.e. as $\varepsilon_j \to 0$. On the other hand, by the Lebesgue dominated convergence theorem, $I_\varepsilon \to (\hat{f})^{\vee}$ for all x in $\mathbb{R}^n$. Therefore $(\hat{f})^{\vee} = f$ a.e. Since $f \in \mathcal{S}$, it follows from continuity that

$$(\hat{f})^{\vee}(x) = f(x), \quad x \in \mathbb{R}^n.$$

$\square$

Corollary 4.9 *Let $f \in \mathcal{S}$. Then $(\hat{f})^{\wedge} = \tilde{f}$, where*

$$\tilde{f}(x) = f(-x), \quad x \in \mathbb{R}^n.$$

Proof Using the Fourier inversion formula, we get

$$(\hat{f})^{\wedge}(x) = (2\pi)^{-n/2} \int_{\mathbb{R}^n} e^{-ix\cdot\xi} \hat{f}(\xi)\, d\xi = (\hat{f})^{\vee}(-x) = f(-x) = \tilde{f}(x)$$

for all x in $\mathbb{R}^n$.

$\square$

Corollary 4.10 $\mathcal{F} : \mathcal{S} \to \mathcal{S}$ *is a bijection.*

Proof By the Fourier inversion formula, we get for all φ in $\mathcal{S}$,

$$\hat{f} = 0 \Rightarrow (\hat{f})^{\vee} = 0 \Rightarrow f = 0.$$

Therefore, $\mathcal{F}: \mathcal{S} \to \mathcal{S}$ is injective. Let $g \in \mathcal{S}$. We need to find a function f in $\mathcal{S}$ such that $\hat{f} = g$. To this end, let $f = \check{g} = \tilde{\hat{g}}$. Then $\hat{f} = \hat{\tilde{\hat{g}}}$. But for all functions φ in $\mathcal{S}$,

$$
\begin{aligned}
\hat{\tilde{\varphi}}(\xi) &= (2\pi)^{-n/2} \int_{\mathbb{R}^n} e^{-ix\cdot\xi} \tilde{\varphi}(x)\, dx = (2\pi)^{-n/2} \int_{\mathbb{R}^n} e^{-ix\cdot\xi} \varphi(-x)\, dx \\
&= (2\pi)^{-n/2} \int_{\mathbb{R}^n} e^{ix\cdot\xi} \varphi(x)\, dx = \check{\varphi}(\xi)
\end{aligned}
$$

for all ξ in $\mathbb{R}^n$. Hence, by the Fourier inversion formula,

$$
\hat{f} = \tilde{\hat{\check{g}}} = g.
$$

So, $\mathcal{F}: \mathcal{S} \to \mathcal{S}$ is surjective. $\qquad\square$

We can now give another important property of the Fourier transform. It is known as the Plancherel theorem.

Theorem 4.11 *Let $\varphi \in \mathcal{S}$. Then $\|\hat{\varphi}\|_2 = \|\varphi\|_2$.*

Proof Let ψ be the function on $\mathbb{R}^n$ defined by

$$
\psi(x) = \overline{\varphi(-x)}, \quad x \in \mathbb{R}^n.
$$

Then

$$
\begin{aligned}
\hat{\psi}(\xi) &= (2\pi)^{-n/2} \int_{\mathbb{R}^n} e^{-ix\cdot\xi} \overline{\varphi(-x)}\, dx \\
&= (2\pi)^{-n/2} \int_{\mathbb{R}^n} e^{ix\cdot\xi} \overline{\varphi(x)}\, dx = \overline{\hat{\varphi}(\xi)}
\end{aligned}
$$

for all ξ in $\mathbb{R}^n$. Thus, by Proposition 4.7 and Corollary 4.9, we get

$$
\begin{aligned}
\|\varphi\|_2^2 &= \int_{\mathbb{R}^n} \hat{\varphi}(\xi)\overline{\hat{\varphi}(\xi)}\, d\xi = \int_{\mathbb{R}^n} \hat{\varphi}(\xi)\hat{\psi}(\xi)\, d\xi \\
&= \int_{\mathbb{R}^n} \hat{\varphi}(\xi)\psi(\xi)\, d\xi = \int_{\mathbb{R}^n} \check{\varphi}(\xi)\psi(\xi)\, d\xi \\
&= \int_{\mathbb{R}^n} \varphi(-\xi)\overline{\varphi(-\xi)}\, d\xi = \int_{\mathbb{R}^n} \varphi(\xi)\overline{\varphi(\xi)}\, d\xi = \|\varphi\|_2^2.
\end{aligned}
$$

$\qquad\square$

Theorem 4.11 allows us to define the Fourier transform of a function in $L^2(\mathbb{R}^n)$ by means of a standard argument in functional analysis. To wit, let $f \in L^2(\mathbb{R}^n)$. Then there exists a sequence $\{\varphi_j\}_{j=1}^\infty$ in $\mathcal{S}$ such that

$$
\varphi_j \to f
$$

in $L^2(\mathbb{R}^n)$ as $j \to \infty$. By Theorem 4.11,

$$
\|\widehat{\varphi_j}\|_2 = \|\varphi_j\|_2, \quad j = 1, 2, \dots.
$$

So, $\{\widehat{\varphi_j}\}_{j=1}^{\infty}$ is a Cauchy sequence in $L^2(\mathbb{R}^n)$ and hence converges to some function, which we denote by $\hat{f}$ in $L^2(\mathbb{R}^n)$. That $\hat{f}$ is independent of the choice of the sequence $\{\varphi_j\}_{j=1}^{\infty}$ is easy to prove. We call $\hat{f}$ the Fourier transform of f. The following result is also known as the Plancherel formula.

Theorem 4.12 *Let $f \in L^2(\mathbb{R}^n)$. Then*

$$\|\hat{f}\|_2 = \|f\|_2.$$

Proof Let $f \in L^2(\mathbb{R}^n)$. Then there exists a sequence $\{\varphi_j\}_{j=1}^{\infty}$ in $\mathcal{S}$ such that

$$\varphi_j \to f$$

in $L^2(\mathbb{R}^n)$ as $j \to \infty$. Then

$$\widehat{\varphi_j} \to \hat{f}$$

in $L^2(\mathbb{R}^n)$ as $j \to \infty$. Therefore

$$\|\hat{f}\|_2 = \lim_{j\to\infty} \|\widehat{\varphi_j}\|_2 = \lim_{j\to\infty} \|\varphi_j\|_2 = \|f\|_2.$$

$\square$

Historical Notes

Jean Baptiste Joseph Fourier (1768–1830) was a French mathematical physicist. He is unquestionably immortal for his work in Fourier series that has blossomed into a mainstream branch of beautiful mathematics known as harmonic analysis today. The series that bear his name were used extensively in his famous book *Théorie Analytique de la Chaleur* published in 1822. In his capacity as prefect of the district of Isère in southeastern France, he built the first real road from Grenoble to Torino.

Bernhard Riemann (1826–1866), a German mathematician, made great contributions to mathematics. His works in analysis and geometry include the Riemannian geometry upon which general relativity is based and the distribution of prime numbers via the Riemann zeta function. The Riemann hypothesis on the location of all the nontrivial zeros of the Riemann zeta function on the vertical line $\operatorname{Re} z = \frac{1}{2}$ in the complex plane is best known as the greatest unsolved problem in mathematics today.

Michel Plancherel (1885–1967) was a Swiss mathematician. The Plancherel theorem stated in this chapter is a special case of the much more sophisticated Plancherel theorems in harmonic analysis and signal analysis, which are still intensively studied nowadays. In the case of classical Fourier series and expansions in terms of orthonormal bases in Hilbert spaces, the Plancherel theorem is often called the Parseval identity.

Exercises

1. Let f be a nonnegative function in $L^1(\mathbb{R}^n)$. Prove that

$$|\hat{f}(\xi)| \le \hat{f}(0), \quad \xi \in \mathbb{R}^n.$$

2. Let φ be the function on $\mathbb{R}^n$ defined by

$$\varphi(x) = e^{-|x|^2/2}, \quad x \in \mathbb{R}^n.$$

   Compute $(\varphi * \varphi)(x)$ for all x in $\mathbb{R}^n$. (Hint: Why is an exercise on convolutions placed in the chapter on Fourier transforms?)

3. Prove that

$$\widehat{\Delta\varphi}(\xi) = -|\xi|^2 \hat{\varphi}(\xi), \quad \xi \in \mathbb{R}^n,$$

   for all φ in $\mathcal{S}$, where

$$\Delta = \sum_{j=1}^{n} \frac{\partial^2}{\partial x_j^2}.$$

4. Prove that for all f and g in $L^2(\mathbb{R}^n)$,

$$(f, g) = \frac{1}{4}(\|f + g\|_2^2 - \|f - g\|_2^2 + i\|f + ig\|_2^2 - i\|f - ig\|_2^2).$$

   (This is called the polarization identity.)

5. Prove that for all functions f and g in $L^2(\mathbb{R}^n)$,

$$(\hat{f}, \hat{g}) = (f, g).$$

   This result gives the most general form of the Plancherel formula.

6. Prove that for all functions f and g in $L^2(\mathbb{R}^n)$,

$$\widehat{f * g} = (2\pi)^{n/2} \hat{f}\hat{g}.$$

7. Is the convolution of two functions in $L^2(\mathbb{R}^n)$ a function in $L^2(\mathbb{R}^n)$?

Chapter 5

Tempered Distributions

The theory of distributions is the language for a serious study of partial differential equations. Only tempered distributions are used in this book. The very minimal amount of distribution theory and the Dirac delta function given in this chapter can be found in the book [58]. More comprehensive accounts are [22, 34, 46].

We first introduce the notion of convergence in the Schwartz space $\mathcal{S}$.

Definition 5.1 Let $\{\varphi_j\}_{j=1}^\infty$ be a sequence of functions in $\mathcal{S}$. Suppose that for all multi-indices α and β,

$$\sup_{x \in \mathbb{R}^n} |x^\alpha(\partial^\beta \varphi_j)(x)| \to 0$$

as $j \to \infty$. Then we say that $\{\varphi_j\}_{j=1}^\infty$ converges to 0 in $\mathcal{S}$ and we sometimes write $\varphi_j \to 0$ in $\mathcal{S}$ as $j \to \infty$.

By the Fourier inversion formula, we know that $\mathcal{F} : \mathcal{S} \to \mathcal{S}$ is a bijection. In fact, we can say a lot more about this bijection.

Theorem 5.2 $\mathcal{F} : \mathcal{S} \to \mathcal{S}$ *is a homeomorphism. This means that* $\mathcal{F} : \mathcal{S} \to \mathcal{S}$ *is a bijection such that* $\mathcal{F} : \mathcal{S} \to \mathcal{S}$ *and* $\mathcal{F}^{-1} : \mathcal{S} \to \mathcal{S}$ *are continuous in the sense that they map convergent sequences in* $\mathcal{S}$ *to convergent sequences in* $\mathcal{S}$.

Proof Let $\{\varphi_j\}_{j=1}^\infty$ be a sequence in $\mathcal{S}$ such that $\varphi_j \to 0$ in $\mathcal{S}$ as $j \to \infty$. Then for all multi-indices α and β,

$$\sup_{\xi \in \mathbb{R}^n} |\xi^\alpha(D^\beta \widehat{\varphi_j})(\xi)| = \sup_{\xi \in \mathbb{R}^n} |\xi^\alpha((-x)^\beta \varphi_j)^\wedge(\xi)| = \sup_{\xi \in \mathbb{R}^n} |\{D^\alpha((-x)^\beta \varphi_j)\}^\wedge(\xi)|$$

$$\leq (2\pi)^{-n/2} \|D^\alpha((-x)^\beta \varphi_j)\|_1.$$

Since $\varphi_j \to 0$ in $\mathcal{S}$ as $j \to \infty$, it follows that for every positive integer N,

$$\sup_{x \in \mathbb{R}^n} \{(1 + |x|)^N |\{D^\alpha((-x)^\beta \varphi_j)\}(x)|\} \to 0$$

DOI: 10.1201/9781003206781-5

as $j \to \infty$. Therefore for every positive integer N with $N > n$,

$$
\begin{aligned}
& \|D^\alpha((-x)^\beta \varphi_j)\|_1 \\
&= \int_{\mathbb{R}^n} |\{D^\alpha((-x)^\beta \varphi_j)\}(x)| \, dx \\
&= \int_{\mathbb{R}^n} (1+|x|)^N |\{D^\alpha((-x)^\beta \varphi_j)\}(x)| \frac{1}{(1+|x|)^N} \, dx \\
&\leq \sup_{x \in \mathbb{R}^n} \{(1+|x|)^N |\{D^\alpha((-x)^\beta \varphi_j)\}(x)|\} \int_{\mathbb{R}^n} (1+|x|)^{-N} \, dx \to 0
\end{aligned}
$$

as $j \to \infty$. Thus,

$$
\sup_{\xi \in \mathbb{R}^n} |\xi^\alpha (D^\beta \widehat{\varphi_j})(\xi)| \to 0
$$

as $j \to \infty$. This proves that $\mathcal{F} : \mathcal{S} \to \mathcal{S}$ is continuous. Moreover, for all φ in $\mathcal{S}$,

$$
\check{\varphi} = \hat{\hat{\varphi}}, \quad \varphi \in \mathcal{S}.
$$

In other words, the inverse Fourier transform is the composition of the Fourier transform and reflection. Since reflection is trivially a continuous mapping of $\mathcal{S}$ into $\mathcal{S}$, it follows that the inverse Fourier transform is also a continuous mapping of $\mathcal{S}$ into $\mathcal{S}$. $\qquad\Box$

We can now come to the notion of a tempered distribution.

Definition 5.3 Let $T : \mathcal{S} \to \mathbb{C}$ be a continuous linear functional. Then we call T a tempered distribution.

Thus, a linear functional $T : \mathcal{S} \to \mathbb{C}$ is a tempered distribution if for every sequence $\{\varphi_j\}_{j=1}^\infty$ of functions in $\mathcal{S}$ with $\varphi_j \to 0$ in $\mathcal{S}$ as $j \to \infty$,

$$
T(\varphi_j) \to 0
$$

as $j \to \infty$. The set of all tempered distributions is denoted by $\mathcal{S}'$.

An important class of tempered distributions comes from the class of tempered functions, which we now introduce.

Definition 5.4 Let f be a measurable function on $\mathbb{R}^n$ such that

$$
\int_{\mathbb{R}^n} \frac{|f(x)|}{(1+|x|)^N} \, dx < \infty
$$

for some positive integer N. Then we call f a tempered function.

Remark 5.5 Functions in $L^p(\mathbb{R}^n)$, $1 \leq p \leq \infty$, are tempered functions. Indeed, let $f \in L^\infty(\mathbb{R}^n)$. Then for every positive integer $N > n$,

$$
\int_{\mathbb{R}^n} \frac{|f(x)|}{(1+|x|)^N} \, dx \leq \|f\|_\infty \int_{\mathbb{R}^n} (1+|x|)^{-N} \, dx < \infty.
$$

Therefore, f is a tempered function. Next, let $f \in L^1(\mathbb{R}^n)$. Then for every positive integer N,

$$\int_{\mathbb{R}^n} \frac{|f(x)|}{(1+|x|)^N} dx \leq \int_{\mathbb{R}^n} |f(x)| \, dx < \infty.$$

So, f is a tempered function. Finally, let $f \in L^p(\mathbb{R}^n)$, $1 < p < \infty$. Then let N be a positive integer such that $Np' > n$, where p' is the conjugate index of p. We have

$$\int_{\mathbb{R}^n} \frac{|f(x)|}{(1+|x|)^N} dx \leq \left(\int_{\mathbb{R}^n} |f(x)|^p dx \right)^{1/p} \left(\int_{\mathbb{R}^n} (1+|x|)^{-Np'} dx \right)^{1/p'} < \infty.$$

Thus, f is a tempered function.

Proposition 5.6 *Let f be a tempered function on $\mathbb{R}^n$. Then the linear functional $T_f : \mathcal{S} \to \mathbb{C}$ defined by*

$$T_f(\varphi) = \int_{\mathbb{R}^n} f(x)\varphi(x) \, dx, \quad \varphi \in \mathcal{S},$$

is a tempered distribution.

Proof Let us begin by proving that the integral exists. Indeed, let N be a positive integer such that

$$\int_{\mathbb{R}^n} \frac{|f(x)|}{(1+|x|)^N} dx < \infty.$$

Then

$$
\begin{aligned}
\int_{\mathbb{R}^n} |f(x)| \, |\varphi(x)| \, dx &= \int_{\mathbb{R}^n} \frac{|f(x)|}{(1+|x|)^N} (1+|x|)^N |\varphi(x)| \, dx \\
&\leq \sup_{x \in \mathbb{R}^n} \{(1+|x|)^N |\varphi(x)|\} \int_{\mathbb{R}^n} \frac{|f(x)|}{(1+|x|)^N} dx < \infty.
\end{aligned}
$$

To prove that $T_f : \mathcal{S} \to \mathbb{C}$ is continuous, let $\{\varphi_j\}_{j=1}^{\infty}$ be a sequence of functions in $\mathcal{S}$ such that $\varphi_j \to$ in $\mathcal{S}$ as $j \to \infty$. Then

$$\sup_{x \in \mathbb{R}^n} \{(1+|x|)^N |\varphi_j(x)|\} \to 0$$

as $j \to \infty$. Therefore

$$
\begin{aligned}
|T_f(\varphi_j)| &\leq \int_{\mathbb{R}^n} |f(x)| \, |\varphi_j(x)| \, dx \\
&\leq \sup_{x \in \mathbb{R}^n} \{(1+|x|)^N |\varphi_j(x)|\} \int_{\mathbb{R}^n} \frac{|f(x)|}{(1+|x|)^N} dx \to 0
\end{aligned}
$$

as $j \to \infty$.   □

We follow the common abuse of terminology and notation in mathematics by identifying the tempered distribution T_f with the tempered function f and writing

$$T_f = f.$$

The following example shows that there are tempered distributions that are not tempered functions.

Example 5.7 Let $\delta : \mathcal{S} \to \mathbb{C}$ be the linear functional defined by

$$\delta(\varphi) = \varphi(0), \quad \varphi \in \mathcal{S}.$$

Then δ is a tempered distribution, but not a tempered function.

Proof Let $\{\varphi_j\}_{j=1}^{\infty}$ be a sequence of functions in $\mathcal{S}$ such that $\varphi_j \to 0$ as $j \to \infty$. Then it is obvious that

$$\delta(\varphi_j) = \varphi_j(0) \to 0$$

as $j \to \infty$. This completes the easy proof that δ is a tempered distribution. In order to prove that δ is not a tempered function, we suppose that $\delta = T_f$ for some tempered function f on $\mathbb{R}^n$. Then

$$\int_{\mathbb{R}^n} f(x)\varphi(x)\, dx = \delta(\varphi) = \varphi(0), \quad \varphi \in \mathcal{S}.$$

For every positive number ε, let $\varphi_\varepsilon \in C_0^{\infty}(\mathbb{R}^n)$ be such that

$$0 \leq \varphi_\varepsilon(x) \leq 1, \quad x \in \mathbb{R}^n,$$

$$\varphi_\varepsilon(0) = 1,$$

and

$$\mathrm{supp}(\varphi_\varepsilon) \subseteq \{x \in \mathbb{R}^n : |x| \leq \varepsilon\}.$$

That such a function φ_ε exists is Exercise 4 in Chapter 3. Then for every nonzero x in $\mathbb{R}^n$, $\varphi_\varepsilon(x) \to 0$ as $\varepsilon \to 0$. Moreover, for $\varepsilon \in (0,1)$, and for every positive integer N with $N > n$,

$$
\begin{aligned}
|f(x)\varphi_\varepsilon(x)| &= \frac{|f(x)|}{(1+|x|)^N}(1+|x|)^N|\varphi_\varepsilon(x)| \\
&\leq \frac{|f(x)|}{(1+|x|)^N}2^N, \quad x \in \mathbb{R}^n.
\end{aligned}
$$

Since $\int_{\mathbb{R}^n} \frac{|f(x)|}{(1+|x|)^N}\, dx < \infty$, it follows from Lebesgue's dominated convergence theorem that

$$\int_{\mathbb{R}^n} f(x)\varphi_\varepsilon(x)\, dx \to 0$$

as $\varepsilon \to 0$. This is a contradiction because

$$\int_{\mathbb{R}^n} f(x)\varphi_\varepsilon(x)\,dx = 1$$

for every positive number ε. □

Remark 5.8 The tempered distribution δ is known as the Dirac delta. We have just shown that it is not a tempered function. In fact, it is not given by any measurable function on $\mathbb{R}^n$, which is integrable on compact subsets of $\mathbb{R}^n$. Of course, Dirac was well aware of this fact. But every time he used δ as if it were a function in his calculations in quantum mechanics, he got the correct answer that could be verified experimentally. So, there must be some truth in this mathematical fiction. It was not until 1950 that Laurent Schwartz removed all the mysteries of the Dirac delta. It is now understood that the Dirac delta is in fact a tempered distribution. Notwithstanding the achievements of Laurent Schwartz, we occasionally follow the practice of Dirac and use the Dirac delta freely as a function in this book. It is both amazing and satisfying to see that it works so well.

The following formula that we shall find very useful presumes that δ is a function.

Theorem 5.9 *For all functions φ in $\mathcal{S}$,*

$$\int_{\mathbb{R}^n} \delta(x - y)\varphi(y)\,dy = \varphi(x), \quad x \in \mathbb{R}^n.$$

The formula does not make sense as it stands. To make sense out of the formula, we need to define the convolution of a tempered distribution T and a Schwartz function φ. To wit, if T is a tempered function, then for all x in $\mathbb{R}^n$,

$$
\begin{aligned}
(T * \varphi)(x) &= \int_{\mathbb{R}^n} T(x - y)\varphi(y)\,dy \\
&= \int_{\mathbb{R}^n} T(y)\varphi(x - y)\,dy \\
&= \int_{\mathbb{R}^n} T(y)\tilde{\varphi}_{-x}(y)\,dy \\
&= T(\tilde{\varphi}_{-x}).
\end{aligned}
$$

Therefore, it is completely reasonable to define for all tempered distributions T, the function $T * \varphi$ on $\mathbb{R}^n$ by

$$(T * \varphi)(x) = T((\tilde{\varphi})_{-x}), \quad x \in \mathbb{R}^n.$$

Theorem 5.9, which is Dirac's way of using the "function" δ, is just the symbolic, and more transparent, way of expressing the following idea.

Theorem 5.10 $\delta * \varphi = \varphi, \quad \varphi \in \mathcal{S}.$

Proof Let $\varphi \in \mathcal{S}$. Then for all x in $\mathbb{R}^n$,

$$(\delta * \varphi)(x) = \delta((\tilde{\varphi})_{-x}) = (\tilde{\varphi})_{-x}(0) = \tilde{\varphi}(-x) = \tilde{\varphi}(-x) = \varphi(x).$$

$\square$

Remark 5.11 Let T be a tempered distribution. Then for all multi-indices α, we define $\partial^\alpha T$ to be the linear functional on $\mathcal{S}$ by

$$(\partial^\alpha T)(\varphi) = (-1)^{|\alpha|} T(\partial^\alpha \varphi), \quad \varphi \in \mathcal{S}.$$

Then it is easy to prove and we leave it as an exercise that $\partial^\alpha T$ is also a tempered distribution. We state without proof the important regularity theorem that every tempered distribution is a derivative of a continuous and tempered function on $\mathbb{R}^n$. (A proof can be found in [34, 40].) Using this regularity theorem, we can prove that the convolution $T * \varphi$ of a tempered distribution T and a Schwartz function φ is a tempered function. Indeed, write

$$T = \partial^\alpha f,$$

where α is a multi-index and f is a continuous and tempered function on $\mathbb{R}^n$. Then for all x in $\mathbb{R}^n$,

$$
\begin{aligned}
(T * \varphi)(x) &= ((\partial^\alpha f) * \varphi)(x) \\
&= (\partial^\alpha f)((\tilde{\varphi})_{-x}) \\
&= (-1)^{|\alpha|} f(\partial^\alpha((\tilde{\varphi})_{-x})) \\
&= (-1)^{|\alpha|} \int_{\mathbb{R}^n} f(y) \partial_y^\alpha(\varphi(x-y)) \, dy \\
&= \int_{\mathbb{R}^n} f(y)(\partial^\alpha \varphi)(x-y) \, dy \\
&= \int_{\mathbb{R}^n} f(x-y)(\partial^\alpha \varphi)(y) \, dy.
\end{aligned}
$$

Let N be a positive integer such that

$$\int_{\mathbb{R}^n} \frac{|f(x)|}{(1+|x|)^N} \, dx < \infty.$$

Then for all x in $\mathbb{R}^n$,

$$
\begin{aligned}
|(T * \varphi)(x)| &\leq \int_{\mathbb{R}^n} \frac{|f(x-y)|}{(1+|x-y|)^N}(1+|x-y|)^N |(\partial^\alpha \varphi)(y)| \, dy \\
&\leq (1+|x|)^N \sup_{y \in \mathbb{R}^n} \{(1+|y|)^N |(\partial^\alpha \varphi)(y)|\} \int_{\mathbb{R}^n} \frac{|f(y)|}{(1+|y|)^N} \, dy
\end{aligned}
$$

and hence $T * \varphi$ is a tempered function.

We end this chapter with a study of Fourier transforms of tempered distributions.

Definition 5.12 Let T be a tempered disrtibution. Then we define the Fourier transform $\hat{T}$ of T to be the linear functional on $\mathcal{S}$ given by

$$\hat{T}(\varphi) = T(\hat{\varphi}), \quad \varphi \in \mathcal{S}.$$

Proposition 5.13 $\hat{T}$ *is also a tempered distribution.*

Proof Let $\{\varphi_j\}_{j=1}^{\infty}$ be a sequence of functions in $\mathcal{S}$ such that $\varphi_j \to 0$ in $\mathcal{S}$ as $j \to \infty$. Then, by Theorem 5.2, $\widehat{\varphi_j} \to 0$ in $\mathcal{S}$ as $j \to \infty$. Since T is a tempered distribution, it follows that

$$\hat{T}(\varphi_j) = T(\widehat{\varphi_j}) \to 0$$

as $j \to \infty$. Therefore $\hat{T}$ is a tempered distribution. $\qquad\square$

We have the following Fourier inversion formula for tempered distributions.

Theorem 5.14 *Let T be a tempered distribution. Then*

$$\hat{\hat{T}} = \tilde{T},$$

where $\tilde{T}$ is defined by

$$\tilde{T}(\varphi) = T(\tilde{\varphi}), \quad \varphi \in \mathcal{S}.$$

Proof Let $\varphi \in \mathcal{S}$. Then by the Fourier inversion formula for Schwartz functions, we get

$$\hat{\hat{T}}(\varphi) = \hat{T}(\hat{\varphi}) = T(\hat{\hat{\varphi}}) = T(\tilde{\varphi}) = \tilde{T}(\varphi).$$

$\qquad\square$

Historical Notes

Paul Adrien Maurice Dirac (1902–1984), an English mathematical physicist, obtained his Ph.D. from the University of Cambridge in 1926 and was Lucasian Professor of Mathematics there until 1969. After his retirement from the University of Cambridge, he held a research professorship at Florida State University. He shared a Nobel Prize in Physics with Erwin Schrödinger in 1933 for their fundamental contributions to quantum mechanics. His treatise titled *The Principles of Quantum Mechanics* published by Oxford University Press is a masterpiece and should be studied by every student seriously interested in quantum mechanics.

Exercises

1. Let α and β be multi-indices. For every $\varphi \in \mathcal{S}$, let $\|\varphi\|_{\alpha,\beta}$ be the number defined by

$$\|\varphi\|_{\alpha,\beta} = \sup_{x \in \mathbb{R}^n} |x^{\alpha}(\partial^{\beta}\varphi)(x)|.$$

Is $\|\ \|_{\alpha,\beta}$ a norm in $\mathcal{S}$?

2. Let H be the function defined on $\mathbb{R}$ by

$$H(x) = \begin{cases} 1, & x > 0, \\ 0, & x \leq 0. \end{cases}$$

What is the derivative H' in the sense of distributions?

3. Is the convolution of two tempered functions on $\mathbb{R}^n$ a tempered function on $\mathbb{R}^n$?

4. Is the Fourier transform of a tempered function on $\mathbb{R}^n$ a tempered function on $\mathbb{R}^n$?

5. Let f be a tempered function on $\mathbb{R}^n$ and let φ be a Schwartz function on $\mathbb{R}^n$. Is $f * \varphi$ a Schwartz function on $\mathbb{R}^n$?

6. Let T be a tempered distribution. Then for all multi-indices α, we define $\partial^\alpha T$ to be the linear functional on $\mathcal{S}$ by

$$(\partial^\alpha T)(\varphi) = (-1)^{|\alpha|} T(\partial^\alpha \varphi), \quad \varphi \in \mathcal{S}.$$

Prove that $\partial^\alpha T$ is a tempered distribution.

7. Let α be a multi-index. Compute the Fourier transform of $D^\alpha \delta$.

8. For all x and ξ in $\mathbb{R}^n$, the integral $\int_{\mathbb{R}^n} e^{ix\cdot\xi} d\xi$ is divergent. Prove that

$$(2\pi)^{-n} \int_{\mathbb{R}^n} e^{ix\cdot\xi} d\xi = \delta.$$

9. Let $f(x) = x^\alpha$, $x \in \mathbb{R}^n$, where α is a multi-index. Compute $\hat{f}$.

10. Let f be one of the following functions on $\mathbb{R}$. Compute $\hat{f}$.

    (a) $f(x) = \cos x$, $x \in \mathbb{R}$.
    (b) $f(x) = \sin x$, $x \in \mathbb{R}$.
    (c) $f(x) = e^{iax}$, $x \in \mathbb{R}$, where $a \in \mathbb{R}$.

11. A sequence $\{T_j\}_{j=1}^\infty$ of tempered distributions is said to converge to the tempered distribution T in $\mathcal{S}'$ as $j \to \infty$ and is denoted by

$$T_j \to T$$

in $\mathcal{S}'$ if for all functions φ in $\mathcal{S}$,

$$T_j(\varphi) \to T(\varphi), \quad \varphi \in \mathcal{S}.$$

Prove that

$$\varphi_\varepsilon \to a\delta$$

in $\mathcal{S}'$ as $\varepsilon \to 0$, where $\{\varphi_\varepsilon : \varepsilon > 0\}$ is the Friedrich mollifier introduced in Theorem 3.4.

12. Is there a function e in $L^1(\mathbb{R}^n)$ such that $e * f = f * e = f$ for all f in $L^1(\mathbb{R}^n)$?

13. Let $T \in \mathcal{S}'$. Then for all y in $\mathbb{R}^n$, we define the translation $T_y T$ and the modulation $M_y T$ of T with respect to y by

$$(T_y T)(\varphi) = T(T_{-y}\varphi), \quad \varphi \in \mathcal{S},$$

and

$$(M_y T)(\varphi) = T(M_y \varphi), \quad \varphi \in \mathcal{S},$$

respectively. For all a in $\mathbb{R} \setminus \{0\}$, we define the dilation $D_a T$ of T by

$$(D_a T)(\varphi) = |a|^{-n} T(D_{1/a}\varphi), \quad \varphi \in \mathcal{S}.$$

Prove that for all $T \in \mathcal{S}'$, $y \in \mathbb{R}^n$, and $a \in \mathbb{R} \setminus \{0\}$,

$$(T_y T)^{\wedge} = M_y \hat{T},$$

$$(M_y T)^{\wedge} = T_{-y} \hat{T},$$

and

$$(D_a T)^{\wedge} = |a|^{-n} D_{1/a} \hat{T}.$$

14. Let $T \in \mathcal{S}'$ and let $f \in C^\infty(\mathbb{R}^n)$ be such that for all multi-indices α, $\partial^\alpha f$ is a tempered function. Then we define the linear functional $fT : \mathcal{S} \to \mathbb{C}$ by

$$(fT)(\varphi) = T(f\varphi), \quad \varphi \in \mathcal{S}.$$

Prove that fT is a tempered distribution.

15. Let $T \in \mathcal{S}'$. Prove that for all multi-indices α,

$$(D^\alpha T)^{\wedge} = x^\alpha \hat{T}.$$

16. Let $T \in \mathcal{S}'$. Prove that for all $\varphi \in \mathcal{S}$,

$$(T * \varphi)^{\wedge} = (2\pi)^{n/2} \hat{\varphi} \hat{T}.$$

Chapter 6

The Heat Kernel

We begin with the task of finding a solution $u = u(x,t)$, $x \in \mathbb{R}^n$, $t > 0$, of the initial value problem

$$\begin{cases} \frac{\partial u}{\partial t}(x,t) = (\Delta u)(x,t), & x \in \mathbb{R}^n, t > 0, \\ u(x,0) = f(x), & x \in \mathbb{R}^n, \end{cases} \tag{6.1}$$

where Δ is the Laplacian on $\mathbb{R}^n$ defined by

$$\Delta = \sum_{j=1}^{n} \frac{\partial^2}{\partial x_j^2}$$

and $f \in \mathcal{S}$. The partial differential equation in (6.1) is known as the heat equation. The trick is to take the partial Fourier transform of u with respect to x. If we do this, then we get

$$\begin{cases} \frac{\partial \hat{u}}{\partial t}(\xi,t) + |\xi|^2 \hat{u}(\xi,t) = 0, & \xi \in \mathbb{R}^n, t > 0, \\ \hat{u}(\xi,0) = \hat{f}(\xi), & \xi \in \mathbb{R}^n. \end{cases} \tag{6.2}$$

Thus, from the first equation in (6.2), we get

$$\hat{u}(\xi,t) = C e^{-|\xi|^2 t}, \quad \xi \in \mathbb{R}^n, t > 0,$$

where C is an arbitrary constant, which depends on ξ. Using the initial condition for $\hat{u}(\xi,0)$ in (6.2), we get $C = \hat{f}(\xi)$. Thus,

$$\hat{u}(\xi,t) = e^{-t|\xi|^2} \hat{f}(\xi), \quad \xi \in \mathbb{R}^n, t > 0.$$

If we take the inverse Fourier transform with respect to ξ, then, by the second formula in Proposition 4.5 and the adjoint formula in Proposition 4.7, we get

$$\begin{aligned} u(x,t) &= (2\pi)^{-n/2} \int_{\mathbb{R}^n} e^{ix\cdot\xi} e^{-t|\xi|^2} \hat{f}(\xi) \, d\xi \\ &= (2\pi)^{-n/2} \int_{\mathbb{R}^n} (M_x e^{-t|\cdot|^2})(\xi) \hat{f}(\xi) \, d\xi \\ &= (2\pi)^{-n/2} \int_{\mathbb{R}^n} (T_{-x}(e^{-t|\cdot|^2})^\wedge)(y) \, f(y) \, dy \\ &= (2\pi)^{-n/2} \int_{\mathbb{R}^n} (e^{-t|\cdot|^2})^\wedge(y-x) \, f(y) \, dy \\ &= \int_{\mathbb{R}^n} k_t(x-y) f(y) \, dy, \quad x \in \mathbb{R}^n, t > 0, \end{aligned}$$

DOI: 10.1201/9781003206781-6

where

$$k_t(x) = (2\pi)^{-n} \int_{\mathbb{R}^n} e^{ix\cdot\xi} e^{-t|\xi|^2} d\xi, \quad x \in \mathbb{R}^n, t > 0.$$

So, by Exercise 8 in Chapter 5, $k_0 = \delta$. For $t > 0$,

$$k_t(x) = (2\pi)^{-n} \int_{\mathbb{R}^n} e^{ix\cdot\xi} (D_{\sqrt{2t}}\varphi)(\xi) \, d\xi, \quad x \in \mathbb{R}^n,$$

where

$$\varphi(\xi) = e^{-|\xi|^2/2}, \quad \xi \in \mathbb{R}^n.$$

By the third formula in Proposition 4.5, the formula for the Fourier transform of the φ in Proposition 4.6 and the adjoint formula in Proposition 4.7, we get

$$k_t(x) = (2\pi)^{-n/2}(2t)^{-n/2} e^{-|x|^2/(4t)} = (4\pi t)^{-n/2} e^{-|x|^2/(4t)}$$

for all x in $\mathbb{R}^n$ and $t > 0$. Therefore the solution u of the initial value problem (6.1) is given by

$$u(x,t) = (k_t * f)(x) = (4\pi t)^{-n/2} \int_{\mathbb{R}^n} e^{-|x-y|^2/(4t)} f(y) \, dy$$

for all x in $\mathbb{R}^n$ and $t > 0$.

We call the function k_t, $t > 0$, the heat kernel or the Weierstrass kernel of the Laplacian Δ on $\mathbb{R}^n$. We also denote $k_t * f$ by $e^{t\Delta} f$ for $t > 0$. The operator $e^{t\Delta}$, $t > 0$, is known as the heat semigroup for the Laplacian.

That $u(x,t) = (k_t * f)(x)$, $x \in \mathbb{R}^n$, $t > 0$, is a solution of the partial differential equation in (6.1) is easy to check. In general, we cannot get the initial condition in (6.1) by simply putting $t = 0$ in the formula for $u(x,t)$. In fact, the initial condition in (6.1) must be suitably interpreted. We make this statement precise in the following two theorems.

Theorem 6.1 *Let $f \in \mathcal{S}$. Then*

$$u(x,0) = f(x), \quad x \in \mathbb{R}^n.$$

Proof Since $k_0 = \delta$, it follows from Theorem 5.10 that

$$u(x,0) = (k_0 * f)(x) = (\delta * f)(x) = f(x), \quad x \in \mathbb{R}^n.$$

$\square$

Theorem 6.2 *Let $f \in L^p(\mathbb{R}^n)$, $1 \le p < \infty$. Then*

$$k_t * f \to f$$

in $L^p(\mathbb{R}^n)$ as $t \to 0+$. If f is a bounded and continuous function on $\mathbb{R}^n$, then

$$k_t * f \to f$$

uniformly on compact subsets of $\mathbb{R}^n$ as $t \to 0+$.

Proof In fact, for all $t > 0$, k_t is equal to the Friedrich mollifier $\psi_{2\sqrt{t}}$, where

$$\psi(x) = \pi^{-n/2} e^{-|x|^2}, \quad x \in \mathbb{R}^n.$$

Since $\int_{\mathbb{R}^n} e^{-|x|^2} dx = \pi^{n/2}$, it follows from Theorem 3.2 that

$$k_t * f = \psi_{2\sqrt{t}} * f \to f$$

in $L^p(\mathbb{R}^n)$ as $t \to 0+$. If f is a bounded and continuous function on $\mathbb{R}^n$, then, by Theorem 3.12,

$$k_t * f = \psi_{2\sqrt{t}} * f \to f$$

uniformly on compact subsets of $\mathbb{R}^n$ at $t \to 0+$. $\square$

Using the heat kernel, we can give estimates for the L^r norms of the solution u of the initial value problem (6.1) in terms of the L^p norms of the initial data f. Such estimates are known as L^p-L^r estimates. The first tool that comes to mind is Young's inequality. Indeed, if $f \in L^p(\mathbb{R}^n)$, $1 \le p \le \infty$, then

$$\|u(\cdot,t)\|_p = \|k_t * f\|_p \le \|k_t\|_1 \|f\|_p, \quad t > 0.$$

But

$$\|k_t\|_1 = \|\psi_{2\sqrt{t}}\|_1 = \int_{\mathbb{R}^n} \psi(x)\, dx = 1, \quad t > 0.$$

So, for $1 \le p \le \infty$,

$$\|u(\cdot,t)\|_p \le \|f\|_p, \quad t > 0. \tag{6.3}$$

Although the estimate in (6.3) is mathematically correct, it is physically inadequate for the following reason. If we think of $u(x,t)$ as the temperature at the position x and time t, then the physics dictates that the temperature should go to zero as $t \to \infty$. In order to obtain a mathematically correct and physically plausible estimate, we use the following result, which is known as the Riesz–Thorin theorem. A proof can be found in the books [45, 54].

Theorem 6.3 *Let α_1, α_2, β_1, and β_2 be real numbers in $[0,1]$. Suppose that*

$$A : L^{1/\alpha_1}(\mathbb{R}^n) \to L^{1/\beta_1}(\mathbb{R}^n)$$

and

$$A : L^{1/\alpha_2}(\mathbb{R}^n) \to L^{1/\beta_2}(\mathbb{R}^n)$$

are bounded linear operators such that there exist positive constants M_1 and M_2 for which

$$\|Af\|_{1/\beta_1} \le M_1 \|f\|_{1/\alpha_1}, \quad f \in L^{1/\alpha_1}(\mathbb{R}^n),$$

and

$$\|Af\|_{1/\beta_2} \le M_2 \|f\|_{1/\alpha_2}, \quad f \in L^{1/\alpha_2}(\mathbb{R}^n).$$

If, for $0 < \theta < 1$, we let

$$\alpha = \theta \alpha_1 + (1 - \theta) \alpha_2$$

and

$$\beta = \theta \beta_1 + (1 - \theta) \beta_2,$$

then $A : L^{1/\alpha}(\mathbb{R}^n) \to L^{1/\beta}(\mathbb{R}^n)$ is a bounded linear operator and

$$\|Af\|_{1/\beta} \leq M_1^\theta M_2^{1-\theta} \|f\|_{1/\alpha}, \quad f \in L^{1/\alpha}(\mathbb{R}^n).$$

We can now give a mathematically correct and physically plausible estimate for the L^∞ norm of the solution u of the initial value problem (6.1) in terms of the L^p norm of f for $1 \leq p < \infty$.

Theorem 6.4 *Let $f \in L^p(\mathbb{R}^n)$, $1 \leq p < \infty$. Let*

$$u(x, t) = (k_t * f)(x), \quad x \in \mathbb{R}^n, t > 0.$$

Then

$$\|u(\cdot, t)\|_\infty \leq (4\pi t)^{-n/(2p)} \|f\|_p, \quad t > 0.$$

Proof Obviously, for $t > 0$,

$$\|k_t * f\|_\infty \leq (4\pi t)^{-n/2} \|f\|_1, \quad f \in L^1(\mathbb{R}^n).$$

By Young's inequality in Theorem 3.2, we get for $t > 0$,

$$\|k_t * f\|_\infty \leq \|k_t\|_1 \|f\|_\infty = \|\psi_{2\sqrt{t}}\|_1 \|f\|_\infty = \|f\|_\infty, \quad f \in L^\infty(\mathbb{R}^n).$$

Let $\theta = 1/p$. Then for

$$\alpha_1 = 1, \ \beta_1 = 0, \ M_1 = (4\pi t)^{-n/2},$$

and

$$\alpha_2 = 0, \ \beta_2 = 0, \ M_2 = 1,$$

we get, by the Riesz–Thorin theorem,

$$\|k_t * f\|_\infty \leq (4\pi t)^{-n/(2p)} \|f\|_p, \quad t > 0.$$

$\square$

Theorem 6.4 is an L^p-L^∞ estimate. If we use the Riesz–Thorin theorem differently, then we can get a different estimate.

Theorem 6.5 *Let $f \in L^1(\mathbb{R}^n)$. Let*

$$u(x, t) = (k_t * f)(x), \quad x \in \mathbb{R}^n, t > 0.$$

Then for $1 \leq p < \infty$,

$$\|u(\cdot, t)\|_{p'} \leq (4\pi t)^{-n/(2p)} \|f\|_1.$$

Proof By Young's inequality, we get for $t > 0$,

$$\|k_t * f\|_\infty \leq \|k_t\|_\infty \|f\|_1 \leq (4\pi t)^{-n/2} \|f\|_1, \quad f \in L^1(\mathbb{R}^n),$$

and

$$\|k_t * f\|_1 \leq \|k_t\|_1 \|f\|_1 = \|f\|_1, \quad f \in L^1(\mathbb{R}^n).$$

Let $\theta = 1/p$. Then for

$$\alpha_1 = 1, \ \beta_1 = 0, \ M_1 = (4\pi t)^{-n/2},$$

and

$$\alpha_2 = 1, \ \beta_2 = 1, \ M_2 = 1,$$

we get by the Riesz–Thorin theorem,

$$\|k_t * f\|_{p'} \leq (4\pi t)^{-n/(2p)} \|f\|_1.$$

$\square$

We can also give estimates for the L^r norms, $r > p$, of the solution u in terms of the L^p norms of f. This requires an extension of the Young inequality given in Theorem 3.2.

Theorem 6.6 *Let p, q, and r be real numbers such that $1 \leq p, q \leq \infty$ and*

$$\frac{1}{p} + \frac{1}{q} = 1 + \frac{1}{r}.$$

If $f \in L^p(\mathbb{R}^n)$ and $g \in L^q(\mathbb{R}^n)$, then $f * g \in L^r(\mathbb{R}^n)$ and

$$\|f * g\|_r \leq \|f\|_p \|g\|_q.$$

Proof By Young's inequality in Theorem 3.2, we get

$$\|f * g\|_q \leq \|f\|_1 \|g\|_q, \quad f \in L^1(\mathbb{R}^n).$$

By Hölder's inequality,

$$\|f * g\|_\infty \leq \|f\|_{q'} \|g\|_q, \quad f \in L^{q'}(\mathbb{R}^n).$$

Let $\theta = q/r$. Then for

$$\alpha_1 = 1, \ \beta_1 = 1/q, \ M_1 = \|g\|_q,$$

and

$$\alpha_2 = 1/q', \ \beta_2 = 0, \ M_2 = \|g\|_q,$$

we get, by the Riesz–Thorin theorem,

$$\|f * g\|_r \leq \|f\|_p \|g\|_q, \quad f \in L^p(\mathbb{R}^n).$$

$\square$

We can now use the extended inequality of Young to get the estimates for the L^r norms of the solution of the initial value problem (6.1) in terms of the L^p norms of f.

Theorem 6.7 *Let $f \in L^p(\mathbb{R}^n)$, $1 \leq p < \infty$, and let*

$$u(x,t) = (k_t * f)(x), \ x \in \mathbb{R}^n, \ t > 0.$$

Then for $r > p$,

$$\|u(\cdot,t)\|_r \leq (2\sqrt{t})^{-n/q'}(\pi/q)^{n/(2q)}\pi^{-n/2}\|f\|_p, \quad t > 0,$$

where

$$\frac{1}{p} + \frac{1}{q} = 1 + \frac{1}{r}$$

and $1 < q \leq \infty$.

Proof By the extension of Young's inequality, we get

$$\|u(\cdot,t)\|_r \leq \|\psi_{2\sqrt{t}}\|_q\|f\|_p, \quad t > 0.$$

But for all positive numbers ε,

$$\|\psi_\varepsilon\|_q = (\pi/q)^{n/(2q)}\pi^{-n/2}\varepsilon^{-n/q'}$$

and the proof is complete for $q \neq \infty$. If $q = \infty$, then $p = 1$ and $r = \infty$. Moreover,

$$(\pi/q)^{n/(2q)} \to 1$$

as $q \to \infty$. So,

$$\|u(\cdot,t)\|_\infty \leq (2\sqrt{t})^{-n}\pi^{-n/2}\|f\|_1 = (4\pi t)^{-n/2}\|f\|_1,$$

which is the same as the estimate obtained using the usual Young's inequality. This completes the proof. $\quad\square$

Corollary 6.8 *Let $f \in L^2(\mathbb{R}^n)$ and let*

$$u(x,t) = (k_t * f)(x), \quad x \in \mathbb{R}^n, \ t > 0.$$

Then

$$\|u(\cdot,t)\|_\infty \leq (8\pi t)^{-n/4}\|f\|_2, \quad t > 0.$$

We end this chapter with an analysis of the initial value problem (6.1) when the initial data f is replaced by a tempered distribution. So, we have the initial value problem

$$\begin{cases} \frac{\partial u}{\partial t}(x,t) = (\Delta u)(x,t), & x \in \mathbb{R}^n, \ t > 0, \\ u(\cdot,0) = T, \end{cases} \tag{6.4}$$

where T is a tempered distribution. It is reasonable to expect and we leave it as an exercise to prove that the function u on $\mathbb{R}^n \times (0,\infty)$ defined by

$$u(x,t) = (T * k_t)(x), \quad x \in \mathbb{R}^n, \ t > 0,$$

is a solution of the heat equation in (6.4). The initial condition in (6.4) is fulfilled in the following sense.

Theorem 6.9 *Let $T \in \mathcal{S}'$. Then for all $\varphi \in \mathcal{S}$,*

$$(T * k_t)(\varphi) \to T(\varphi)$$

as $t \to 0+$.

The proof of Theorem 6.9 is based on the following result, which supplements Theorems 3.4 and 3.12.

Theorem 6.10 *Let $\varphi \in \mathcal{S}$ be such that*

$$\int_{\mathbb{R}^n} \varphi(x)\, dx = a.$$

Then for all $f \in \mathcal{S}$,

$$f * \varphi_\varepsilon \to af$$

in $\mathcal{S}$ as $\varepsilon \to 0$, where $\{\varphi_\varepsilon : \varepsilon > 0\}$ is the Friedrich mollifier associated to φ.

Proof By Proposition 4.1 and the third formula in Proposition 4.5, we get for all positive numbers ε,

$$
\begin{aligned}
(f * \varphi_\varepsilon - af)^\wedge(\xi) &= (2\pi)^{n/2}\hat{f}(\xi)\widehat{\varphi_\varepsilon}(\xi) - a\hat{f}(\xi) \\
&= (2\pi)^{n/2}\hat{f}(\xi)\hat{\varphi}(\varepsilon\xi) - a\hat{f}(\xi) \\
&= \{(2\pi)^{n/2}\hat{\varphi}(\varepsilon\xi) - a\}\hat{f}(\xi)
\end{aligned}
$$

for all ξ in $\mathbb{R}^n$. Let α and β be multi-indices. Then, by the Leibniz formula in Proposition 1.1,

$$
\begin{aligned}
&\xi^\alpha(\partial^\beta\{(f * \varphi_\varepsilon - af)^\wedge\})(\xi) \\
&= \xi^\alpha \sum_{\gamma \leq \beta} \binom{\beta}{\gamma} \partial^\gamma\{(2\pi)^{n/2}\hat{\varphi}(\varepsilon\xi) - a\}(\partial^{\beta-\gamma}\hat{f})(\xi) \\
&= \xi^\alpha\{(2\pi)^{n/2}\hat{\varphi}(\varepsilon\xi) - a\}(\partial^\beta\hat{f})(\xi) \\
&\quad + \xi^\alpha \sum_{\gamma \neq 0} \binom{\beta}{\gamma}(2\pi)^{n/2}\varepsilon^{|\gamma|}(\partial^\gamma\hat{\varphi})(\varepsilon\xi)(\partial^{\beta-\gamma}\hat{f})(\xi)
\end{aligned}
$$

for all $\xi \in \mathbb{R}^n$. Obviously,

$$
\sup_{\xi \in \mathbb{R}^n}\left|\xi^\alpha \sum_{\gamma \neq 0} \binom{\beta}{\gamma}(2\pi)^{n/2}\varepsilon^{|\gamma|}(\partial^\gamma\hat{\varphi})(\varepsilon\xi)(\partial^{\beta-\gamma}\hat{f})(\xi)\right| \to 0
$$

as $\varepsilon \to 0$. Let δ be a given positive number. Then there exists a positive number ε_0 such that

$$\varepsilon < \varepsilon_0 \Rightarrow \sup_{\xi \in \mathbb{R}^n}|\xi^\alpha\{(2\pi)^{n/2}\hat{\varphi}(\varepsilon\xi) - a\}(\partial^\beta\hat{f})(\xi)| < \frac{\delta}{2}.$$

Thus,
$$(f * \varphi_\varepsilon - af)^\wedge \to 0$$

in $\mathcal{S}$ as $\varepsilon \to 0$. Since $\mathcal{F} : \mathcal{S} \to \mathcal{S}$ is a homeomorphism, it follows that the proof is complete. □

Proof of Theorem 6.9 By the regularity theorem for tempered distributions, we write
$$T = \partial^\alpha f,$$

where α is a multi-index and f is a continuous and tempered function on $\mathbb{R}^n$. So, for all $t > 0$,

$$
\begin{aligned}
(T * k_t)(x) &= T((\tilde{k}_t)_{-x}) = (\partial^\alpha f)((\tilde{k}_t)_{-x}) \\
&= \int_{\mathbb{R}^n} (\partial^\alpha f)(y) k_t(x - y)\, dy \\
&= \int_{\mathbb{R}^n} f(y)(\partial^\alpha k_t)(x - y)\, dy
\end{aligned}
$$

for all x in $\mathbb{R}^n$. Hence for all $\varphi \in \mathcal{S}$ and $t > 0$, we get

$$
\begin{aligned}
(T * k_t)(\varphi) &= \int_{\mathbb{R}^n} f(y) \left(\int_{\mathbb{R}^n} (\partial^\alpha k_t)(x - y)\varphi(x)\, dx \right) dy \\
&= (-1)^{|\alpha|} \int_{\mathbb{R}^n} f(y) \left(\int_{\mathbb{R}^n} k_t(y - x)(\partial^\alpha \varphi)(x)\, dx \right) dy \\
&= (-1)^{|\alpha|} \int_{\mathbb{R}^n} f(y)(k_t * (\partial^\alpha \varphi))(y)\, dy.
\end{aligned}
$$

By Theorem 6.10,
$$k_t * (\partial^\alpha \varphi) \to \partial^\alpha \varphi$$

in $\mathcal{S}$ as $t \to 0+$. Therefore
$$(T * k_t)(\varphi) \to (-1)^{|\alpha|} \int_{\mathbb{R}^n} f(y)(\partial^\alpha \varphi)(y)\, dy$$

as $t \to 0+$. But
$$(-1)^{|\alpha|} \int_{\mathbb{R}^n} f(y)(\partial^\alpha \varphi)(y)\, dy = T(\varphi)$$

and this completes the proof. □

It is an interesting fact that the singularities of the initial data T in $\mathcal{S}'$ disappear instantaneously as soon as time begins. To be more precise, we have the following result.

Theorem 6.11 *Let $T \in \mathcal{S}'$ and let*
$$u(x, t) = (T * k_t)(x), \quad x \in \mathbb{R}^n, \, t > 0.$$

Then for all $t > 0$,
$$u(\cdot, t) \in C^\infty(\mathbb{R}^n).$$

We leave the proof of Theorem 6.11 as an exercise.

Historical Notes

Pierre Simon de Laplace (1749–1827) was a French mathematician and astronomer. His main scientific interests were in celestial mechanics, probability theory, and related topics in analysis. The partial differential operator Δ that bears his name is familiar to every undergraduate student in mathematics, science, and engineering. His treatises are the five volumes of the *Mécanique Céleste* published from 1799 to 1825 and *Théorie Analytique des Probabilités* published in 1812. The Laplace transform and other tools in analysis due to him are widely used nowadays. The Laplacian, which can be defined on a Riemannian manifold, is a natural partial differential operator to study in order to understand the geometry of the manifold.

Karl Theodor Wilhelm Weierstrass (1815–1897), a German mathematician, taught at a secondary school in Deutsch-Krone, West Prussia, from 1842 to 1848. Then he was a lecturer at Roman Catholic School in Braunsberg. He did mathematical research in his spare time and became famous by publishing a paper on abelian integrals in Crelle's journal. This earned him an honorary doctorate from the University of Königsberg, a professorship in mathematics at the Royal Polytechnic School in Berlin, and an associate professorship at the University of Berlin. His publication in 1871 that there exist continuous functions that have derivatives nowhere is well known to all students in analysis.

Marcel Riesz (1886–1969) was a Hungarian mathematician who spent most of his life at Lund University in Sweden. He was an expert in analysis. The Riesz–Thorin theorem has been extended in various directions to a collection of results known as interpolation theorems in analysis.

G. Olof Thorin (1912–2004) was a Swedish mathematician. He was a student of Marcel Riesz.

Exercises

1. Find the solution of the initial value problem

$$\begin{cases} \frac{\partial u}{\partial t}(x,t) = (\Delta u)(x,t) + au(x,t), & x \in \mathbb{R}^n, t > 0, \\ u(x,0) = f(x), & x \in \mathbb{R}^n, \end{cases}$$

where a is a constant and f is a function in $\mathcal{S}$.

2. Prove that the function u on $\mathbb{R}^n \times (0,\infty)$ defined by

$$u(x,t) = (T * k_t)(x), \quad x \in \mathbb{R}^n, t > 0, \qquad (6.5)$$

satisfies the heat equation in (6.4).

3. Prove that the solution u of the initial value problem (6.4) given by (6.5) is a tempered function in $C^\infty(\mathbb{R}^n)$ for $t > 0$.

4. Use the heat kernel k_t, $t > 0$, to find a solution of the initial value problem

$$\begin{cases} \frac{\partial u}{\partial t}(x,t) = \frac{\partial^2 u}{\partial x^2}(x,t), & x \in \mathbb{R}, \, t > 0, \\ u(x,0) = x, & x \in \mathbb{R}. \end{cases}$$

Does $k_t * f$ converge to f in any sense as $t \to 0+$?

5. Find a solution u of the initial value problem (6.1) when the initial data f is a harmonic function on $\mathbb{R}^n$, i.e.,

$$(\Delta f)(x) = 0, \quad x \in \mathbb{R}^n.$$

6. Let $f \in L^\infty(\mathbb{R}^n)$. Does the solution u of (6.1) given by

$$u(\cdot, t) = k_t * f, \quad t > 0,$$

have to go to zero in any sense as $t \to \infty$?

7. Can an initial value problem

$$\begin{cases} \frac{\partial u}{\partial t}(x,t) = (\Delta u)(x,t), & x \in \mathbb{R}^n, \, t > 0, \\ u(\cdot, 0) = T, \end{cases}$$

where T is a tempered distribution on $\mathbb{R}^n$, have multiple solutions? Explain your answer.

Chapter 7

The Free Propagator

The same technique in the previous chapter can be used to solve the initial value problem for the Schrödinger equation given by

$$\begin{cases} \frac{\partial u}{\partial t}(x,t) = -i(\Delta u)(x,t), & x \in \mathbb{R}^n, \, t \neq 0, \\ u(x,0) = f(x), & x \in \mathbb{R}^n, \end{cases} \tag{7.1}$$

where f is again a function in $\mathcal{S}$.

In spite of the slight modification of the heat equation by multiplying the Laplacian by $-i$, the nature of the equation and the properties of the solution are changed completely. The heat kernel gives the heat flow, while the free propagator that we shall derive describes the quantum mechanical motion of a particle in a vacuum. The solution $u(x,t)$, $x \in \mathbb{R}^n$, $-\infty < t < \infty$, represents the state of the particle at the position x and time t. Hence, the free propagator is also known as the Schrödinger kernel.

In order to solve the initial value problem, we take the partial Fourier transform of the Schrödinger equation with respect to x. Then we get

$$\begin{cases} \frac{\partial \hat{u}}{\partial t}(\xi,t) = i|\xi|^2 \hat{u}(\xi,t), & \xi \in \mathbb{R}^n, \, t \neq 0, \\ \hat{u}(\xi,0) = \hat{f}(\xi), & \xi \in \mathbb{R}^n. \end{cases}$$

Therefore

$$\hat{u}(\xi,t) = \hat{f}(\xi)e^{i|\xi|^2 t}, \quad \xi \in \mathbb{R}^n, \, -\infty < t < \infty,$$

and by taking the inverse Fourier transform with respect to ξ, using the formula for the Fourier transform of the modulation of a tempered distribution in Exercise 13 in Chapter 5, we get

$$\begin{aligned} u(x,t) &= (2\pi)^{-n/2} \int_{\mathbb{R}^n} e^{ix\cdot\xi} e^{i|\xi|^2 t} \hat{f}(\xi) \, d\xi \\ &= (2\pi)^{-n/2} \int_{\mathbb{R}^n} (M_x e^{i|\cdot|^2 t})(\xi) \hat{f}(\xi) \, d\xi \\ &= (2\pi)^{-n/2} \int_{\mathbb{R}^n} (e^{i|\cdot|^2 t})^{\wedge}(y-x) \, f(y) \, dy \\ &= \int_{\mathbb{R}^n} K_t(x-y) \, f(y) \, dy, \quad x \in \mathbb{R}^n, \, -\infty < t < \infty, \end{aligned}$$

DOI: 10.1201/9781003206781-7

where

$$K_t(x) = (2\pi)^{-n} \int_{\mathbb{R}^n} e^{ix\cdot\xi} e^{i|\xi|^2 t} d\xi, \quad x \in \mathbb{R}^n, \ -\infty < t < \infty. \qquad (7.2)$$

Note that by Exercise 8 in Chapter 5, $K_0 = \delta$. To compute K_t for $t \neq 0$, we first assume that $t > 0$. Then for all x in $\mathbb{R}^n$,

$$
\begin{aligned}
K_t(x) &= (2\pi)^{-n} \int_{\mathbb{R}^n} e^{ix\cdot\xi} e^{i|\xi|^2 t} d\xi \\
&= \prod_{j=1}^{n} (2\pi)^{-1} \int_{-\infty}^{\infty} e^{ix_j \xi_j} e^{it\xi_j^2} d\xi_j \\
&= \prod_{j=1}^{n} (2\pi)^{-1} e^{-ix_j^2/(4t)} \int_{-\infty}^{\infty} e^{it\left(\xi_j + \frac{x_j}{2t}\right)^2} d\xi_j \\
&= \prod_{j=1}^{n} (2\pi)^{-1} e^{-ix_j^2/(4t)} t^{-1/2} \int_{-\infty}^{\infty} e^{i\xi_j^2} d\xi_j \\
&= (2\pi)^{-n} t^{-n/2} e^{-i|x|^2/(4t)} \left(\int_{-\infty}^{\infty} e^{i\xi^2} d\xi \right)^n.
\end{aligned}
$$

We now invoke a lemma.

Lemma 7.1 $\int_{-\infty}^{\infty} e^{\pm ix^2} dx = \sqrt{\frac{\pi}{2}}(1 \pm i)$.

Proof We first look at $\int_{-\infty}^{\infty} e^{ix^2} dx$. It is enough to prove that

$$\int_{0}^{\infty} e^{ix^2} dx = \frac{\sqrt{2\pi}}{4}(1 + i).$$

To do this, we integrate e^{iz^2} along the boundary C_ρ of the sector S_ρ given by

$$S_\rho = \left\{ re^{i\theta} : 0 \le \theta \le \frac{\pi}{4}, \ 0 \le r \le \rho \right\}.$$

Therefore

$$\int_{C_\rho} e^{iz^2} dz = \int_{0}^{\rho} e^{ix^2} dx + \int_{0}^{\pi/4} e^{i\rho^2 e^{2i\theta}} i\rho e^{i\theta} d\theta - e^{i\pi/4} \int_{0}^{\rho} e^{-r^2} dr.$$

By Cauchy's integral theorem,

$$\int_{C_\rho} e^{iz^2} dz = 0, \quad \rho > 0.$$

Also, for all $\rho > 0$,

$$\left| \int_{0}^{\pi/4} e^{i\rho^2 e^{2i\theta}} i\rho e^{i\theta} d\theta \right| \le \rho \int_{0}^{\pi/4} e^{-\rho^2 \sin 2\theta} d\theta.$$

By looking at the graphs of $y = \sin\theta$ and $y = 2\theta/\pi$ on $[0, \pi/2]$, we get

$$\sin 2\theta \geq 4\theta/\pi, \quad \theta \in [0, \pi/4].$$

So,

$$\left| \int_0^{\pi/4} e^{i\rho^2 e^{2i\theta}} i\rho\, e^{i\theta}\, d\theta \right| \leq \rho \int_0^{\pi/4} e^{-4\rho^2\theta/\pi}\, d\theta = \frac{\pi}{4\rho}\left(1 - e^{-\rho^2}\right) \to 0$$

as $\rho \to \infty$. Therefore, letting $\rho \to \infty$, we get

$$0 = \int_0^\infty e^{ix^2}\, dx - e^{i\pi/4} \int_0^\infty e^{-r^2}\, dr$$

or

$$\int_0^\infty e^{ix^2}\, dx = \frac{\sqrt{\pi}}{2} e^{i\pi/4} = \frac{\sqrt{\pi}}{2}\left(\frac{\sqrt{2}}{2} + i\frac{\sqrt{2}}{2}\right) = \frac{\sqrt{2\pi}}{4}(1 + i).$$

Therefore

$$\int_{-\infty}^\infty e^{ix^2}\, dx = \sqrt{\frac{\pi}{2}}(1 + i).$$

Using the real part and the imaginary part of $\int_{-\infty}^\infty e^{ix^2}\, dx$, we see that

$$\int_{-\infty}^\infty e^{-ix^2}\, dx = \sqrt{\frac{\pi}{2}}(1 - i).$$

$\square$

With the lemma in place, we get

$$u(x, t) = (K_t * f)(x), \quad x \in \mathbb{R}^n,\, t > 0,$$

where

$$K_t(x) = e^{in\pi/4}(4\pi t)^{-n/2} e^{-i|x|^2/(4t)}, \quad x \in \mathbb{R}^n,\, t > 0.$$

Similarly,

$$u(x, t) = (K_t * f)(x), \quad x \in \mathbb{R}^n,\, t < 0,$$

where

$$K_t(x) = e^{-in\pi/4}(-4\pi t)^{-n/2} e^{-i|x|^2/(4t)}, \quad x \in \mathbb{R}^n,\, t < 0.$$

So, we can write

$$K_t(x) = e^{(\text{sgn } t)in\pi/4}(4\pi|t|)^{-n/2} e^{-i|x|^2/(4t)}, \quad x \in \mathbb{R}^n,\, t \neq 0. \qquad (7.3)$$

We sometimes denote $K_t * f$ by $e^{-i\Delta t} f$ for $-\infty < t < \infty$ and call $e^{-i\Delta t}$, $-\infty < t < \infty$, the free propagator or the one-parameter Schrödinger group.

Let $f \in \mathcal{S}$. If we let u be the function defined by

$$u(x, t) = (K_t * f)(x), \quad x \in \mathbb{R}^n,$$

then we leave as an exercise to prove that u is a solution of the Schrödinger equation in the initial value problem (7.1). As for the initial condition, we have the following theorem.

Theorem 7.2 *Let $f \in \mathcal{S}$. Then*

$$u(x, 0) = f(x), \quad x \in \mathbb{R}^n.$$

Proof Since $K_0 = \delta$, it follows from Theorem 5.10 that

$$u(x, 0) = (K_0 * f)(x) = (\delta * f)(x) = f(x), \quad x \in \mathbb{R}^n.$$

$\square$

In applications, especially in quantum mechanics, the function f in the initial condition is usually in $L^2(\mathbb{R}^n)$. A solution u of the initial value problem (7.1) is formally given by

$$u(\cdot, t) = K_t * f, \quad t \neq 0.$$

What is the meaning of $K_t * f$ when $K_t \in L^\infty(\mathbb{R}^n)$ and $f \in L^2(\mathbb{R}^n)$? A possible meaning can be given in terms of tempered distributions. To see how, let us first assume that f is a very nice function, say, a function in $\mathcal{S}$. Then, by Fubini's theorem and the fact that K_t is a radial function with respect to the spatial variable, we get for all functions φ in $\mathcal{S}$,

$$
\begin{aligned}
(K_t * f)(\varphi) &= \int_{\mathbb{R}^n} (K_t * f)(x)\varphi(x)\,dx \\
&= \int_{\mathbb{R}^n} \left\{ \int_{\mathbb{R}^n} K_t(x - y)f(y)\,dy \right\} \varphi(x)\,dx \\
&= \int_{\mathbb{R}^n} f(y) \left\{ \int_{\mathbb{R}^n} K_t(y - x)\varphi(x)\,dx \right\} dy \\
&= f(K_t * \varphi).
\end{aligned}
$$

So, for every tempered distribution T, we can define $K_t * T : \mathcal{S} \to \mathbb{C}$ by

$$(K_t * T)(\varphi) = T(K_t * \varphi), \quad \varphi \in \mathcal{S}.$$

It is left as an exercise to prove that for $t \neq 0$, $T * K_t$ is a tempered distribution.

It can be proved and is left as an exercise that for all tempered distributions T,

$$(K_t * T)^\wedge = (2\pi)^{n/2}\widehat{K_t}\hat{T} = e^{i|\xi|^2 t}\hat{T}, \quad t \neq 0. \tag{7.4}$$

Now, we can prove the following theorem.

Theorem 7.3 *Let $f \in L^2(\mathbb{R}^n)$. Then the tempered distribution u defined by*

$$u(\cdot, t) = K_t * f, \quad t \neq 0,$$

is a solution in the sense of tempered distributions of the Schrödinger equation in (7.1).

Proof For $t \neq 0$,

$$
\begin{aligned}
\frac{\partial \hat{u}}{\partial t}(\cdot, t) &= i |\cdot|^2 e^{i |\cdot|^2 t} \hat{f} \\
&= e^{i |\cdot|^2 t} (-i \Delta f)^{\wedge} \\
&= (2\pi)^{n/2} \widehat{K_t} (-i \Delta f)^{\wedge} \\
&= (K_t * (-i \Delta f))^{\wedge}.
\end{aligned}
$$

Thus, for $t \neq 0$,

$$
\frac{\partial u}{\partial t}(\cdot, t) = K_t * (-i \Delta f).
$$

But for $t \neq 0$ and for all φ in $\mathcal{S}$,

$$
\begin{aligned}
((K_t * (-i \Delta f)))(\varphi) &= -i(\Delta f)(K_t * f) \\
&= -i f(\Delta(K_t * \varphi)) \\
&= -i f(K_t * \Delta \varphi) \\
&= -i (K_t * f)(\Delta \varphi) \\
&= -i(\Delta(K_t * f))(\varphi) \\
&= -i((\Delta u)(\cdot, t))(\varphi).
\end{aligned}
$$

Therefore for $t \neq 0$,

$$
\frac{\partial u}{\partial t}(\cdot, t) = -i(\Delta u)(\cdot, t)
$$

in the sense of tempered distributions. □

Theorem 7.4 *Let $f \in L^2(\mathbb{R}^n)$. Then*

$$
K_t * f \to f
$$

in $L^2(\mathbb{R}^n)$ as $t \to 0$.

Proof By (7.2) and (7.4), we get for all $t \neq 0$,

$$
(K_t * f)^{\wedge}(\xi) = (2\pi)^{n/2} \widehat{K_t}(\xi) \hat{f}(\xi) = e^{i|\xi|^2 t} \hat{f}(\xi)
$$

for all ξ in $\mathbb{R}^n$. So, by the Plancherel formula in Theorem 4.12,

$$
\|K_t * f - f\|_2^2 = \|(K_t * f)^{\wedge} - \hat{f}\|_2^2 = \int_{\mathbb{R}^n} |e^{i|\xi|^2 t} - 1|^2 |\hat{f}(\xi)|^2 d\xi.
$$

Obviously, for almost all ξ in $\mathbb{R}^n$,

$$
|e^{i|\xi|^2 t} - 1|^2 |\hat{f}(\xi)|^2 \to 0
$$

as $t \to 0$ and

$$
|e^{i|\xi|^2 t} - 1|^2 |\hat{f}(\xi)|^2 \leq 4|\hat{f}(\xi)|^2, \quad \xi \in \mathbb{R}^n.
$$

So, by Lebesgue's dominated convergence theorem,

$$\|K_t * f - f\|_2 \to 0$$

as $t \to 0$ and the proof is complete. $\qquad\qquad\qquad\qquad\qquad\qquad\square$

The free propagator satisfies the law of conservation of energy. This is made precise in the following theorem.

Theorem 7.5 *For all f in $L^2(\mathbb{R}^n)$,*

$$\|e^{-i\Delta t}f\|_2 = \|f\|_2, \quad -\infty < t < \infty.$$

Proof Let $f \in L^2(\mathbb{R}^n)$. Then, by (7.4) and Plancherel's formula in Theorem 4.12,

$$\|e^{-i\Delta t}f\|_2 = \|u(\cdot,t)\|_2 = \|\hat{u}(\cdot,t)\|_2 = \|\hat{f}\|_2 = \|f\|_2$$

for all $t \neq 0$. $\qquad\qquad\qquad\qquad\qquad\qquad\qquad\qquad\qquad\qquad\qquad\square$

Using the law of conservation of energy and the Riesz–Thorin theorem, we can obtain estimates for the solutions of the initial value problem (7.1) for the Schrödinger equation.

Theorem 7.6 *Let $f \in L^p(\mathbb{R}^n)$, $1 \leq p \leq 2$. Then for all $t \neq 0$,*

$$\|e^{-i\Delta t}f\|_{p'} \leq (4\pi|t|)^{-(n/p)+(n/2)}\|f\|_p,$$

where p' is the conjugate index of p.

Proof By Young's inequality in Theorem 3.2, we get for all $t \neq 0$,

$$\|e^{-i\Delta t}f\|_\infty = \|K_t * f\|_\infty \leq \|K_t\|_\infty\|f\|_1 = (4\pi|t|)^{-n/2}\|f\|_1, \quad f \in L^1(\mathbb{R}^n).$$

By the conversation of energy in Theorem 7.5, we get for all $t \neq 0$,

$$\|e^{-i\Delta t}f\|_2 = \|f\|_2, \quad f \in L^2(\mathbb{R}^n).$$

Let $\theta = \frac{2}{p} - 1$. Then for

$$\alpha_1 = 1, \ \beta_1 = 0, \ M_1 = (4\pi|t|)^{-n/2},$$

and

$$\alpha_2 = 1/2, \ \beta_2 = 1/2, \ M_2 = 1,$$

we get, by the Riesz–Thorin theorem, the asserted estimate. $\qquad\qquad\square$

Remark 7.7 Since the heat kernel k_t for $t > 0$ is given by

$$k_t(x) = \frac{1}{(4\pi t)^{n/2}}e^{-|x|^2/(4t)}, \quad x \in \mathbb{R}^n,$$

it is reasonable to guess heuristically that the free propagator K_t for $t \neq 0$ can be obtained by replacing t by $-it$ and hence

$$K_t(x) = k_{-it}(x) = \frac{1}{(4\pi(-it))^{n/2}} e^{-|x|^2/(-4it)} = \frac{1}{(4\pi(-it))^{n/2}} e^{-i|x|^2/(4t)}$$

(7.5)

for all x in $\mathbb{R}^n$. The factor $e^{(\operatorname{sgn} t)in\pi/4}$ comes from the term $(\pm i)^{n/2}$. The idea of obtaining the Schrödinger kernel from the heat kernel can be attributed to Wick. The transformation $t \to -it$ is a technique originating from quantum field theory and is known as the Wick rotation, which transfers computations from the Minkowski space to the Euclidean space $\mathbb{R}^4$.

Remark 7.8 A good account of the free propagator can be found in Chapter 5 of the book [33]. The computation of the inverse Fourier transform of $e^{-it|\xi|^2}$, $\xi \in \mathbb{R}^n$, $t > 0$, is adapted from that used in [50]. The computation of the integral $\int_{-\infty}^{\infty} e^{\pm ix^2} dx$ is a standard exercise in complex analysis. See, for instance, the books [29, 44, 56].

Historical Notes

Guido Fubini (1879–1943) was an Italian mathematician who obtained his doctorate in 1900 from Scuolo Normale Superiorie di Pisa. He taught at Università di Genova, Università di Catania, Politecnico Torino, Università di Torino, and Princeton University. Fubini's theorem on the interchange of the order of integration is well known to every student in mathematical analysis.

Erwin Schrödinger (1887–1961) was an eminent Austrian theoretical physicist who shared a Nobel prize in physics with Paul Adrien Maurice Dirac in 1933 for their contributions to the foundations of quantum mechanics. The Schrödinger equation is a fundamental equation in quantum mechanics and is still at the heart of many active areas of research in modern analysis and mathematical physics.

Augustin-Louis Cauchy (1789–1857), a French mathematician, was born in Paris. He is usually considered the founder of modern analysis, which is a subject in which a high standard of rigor is built into calculus. His name is a household name for every student in analysis nowadays. He was appointed a full professor at École Polytechnique in 1816. He was appointed to the Chair of the Faculty of Science at Collegè de France in 1830. In 1830, Charles X was overthrown by the July revolution, and for some political reasons, Cauchy was forced to resign as Chair. He went into exile and was appointed to the Chair of Mathematical Physics at Università di Torino. He returned to Paris to resume his position at École Polytechnique. He was a professor at Sorbonne from 1848 to 1852.

Gian-Carlo Wick (1909–1992) was born in Torino. He was an Italian theoretical physicist educated in Göttingen and Leipzig, and he spent some time working in the United States. The Wick rotation, first introduced in quantum field theory, has also found applications in the no-boundary theory of the universe through the idea of imaginary time. Wick was a member of the National Academy of Sciences and the Accademia dei Lincei.

Exercises

1. Find a solution of the initial value problem

$$\begin{cases} \frac{\partial u}{\partial t}(x,t) = -i(\Delta u)(x,t), & x \in \mathbb{R}^n, t \neq 0, \\ u(\cdot, 0) = \delta. \end{cases}$$

2. Let $f \in \mathcal{S}$. Prove that for $t \neq 0$, $K_t * f \in \mathcal{S}$.

3. For $t \neq 0$, let

$$u(\cdot, t) = K_t * f,$$

   where $f \in \mathcal{S}$. Prove that

$$\frac{\partial u}{\partial t}(x,t) = -i(\Delta u)(x,t), \quad x \in \mathbb{R}^n, t \neq 0.$$

4. Let T be a tempered distribution. Prove that the linear functional $K_t * T : \mathcal{S} \to \mathbb{C}$ defined by

$$(K_t * T)(\varphi) = T(K_t * \varphi), \quad \varphi \in \mathcal{S},$$

   is a tempered distribution.

5. Let T be a tempered distribution. Prove that for $t \neq 0$,

$$(K_t * T)^\wedge = (2\pi)^{n/2} \widehat{K_t} \hat{T}.$$

6. Let $f \in L^2(\mathbb{R}^n)$. Prove that for $t \neq 0$, $K_t * f \in L^2(\mathbb{R}^n)$.

7. Prove that (7.5) is the same as (7.3) for the Schrödinger kernel K_t, $t \neq 0$.

8. Does the initial value problem (7.1) have a unique solution u for all functions f in $L^2(\mathbb{R}^n)$?

Chapter 8

The Newtonian Potential

Given a function f in $\mathcal{S}$, suppose that we are interested in solving the partial differential equation

$$\Delta u = f \tag{8.1}$$

on $\mathbb{R}^n$ for the unknown u. This is usually called the Poisson equation and is called the Laplace equation when $f = 0$.

It is not unnatural to say that $u = \Delta^{-1} f$ is a solution. What is Δ^{-1}?

To proceed formally, we first observe that, by Exercises 3 and 5 in Chapter 4, Δ is a negative operator in the sense that

$$(\Delta \varphi, \varphi) = (\widehat{\Delta \varphi}, \hat{\varphi}) = -\int_{\mathbb{R}^n} |\xi|^2 |\hat{\varphi}(\xi)|^2 d\xi \le 0$$

for all functions φ in $\mathcal{S}$. Then we take the bold step of treating Δ as a negative number and write

$$\Delta^{-1} = -\int_0^\infty e^{t\Delta} dt.$$

The formula can be made rigorous. But for us it is the attractive formal aspects that convince us that some formula coming out of it should solve the Poisson equation. That the formula is indeed a solution can then be verified mathematically.

Since we know the formula for $e^{t\Delta} f$ for $t > 0$, it follows that for $x \in \mathbb{R}^n$ and $t > 0$,

$$
\begin{aligned}
(\Delta^{-1} f)(x) &= -\left(\left(\int_0^\infty e^{t\Delta} dt\right) f\right)(x) = -\int_0^\infty (e^{t\Delta} f)(x)\, dt \\
&= -\int_0^\infty \int_{\mathbb{R}^n} k_t(x - y) f(y)\, dy\, dt \\
&= -\int_{\mathbb{R}^n} \left(\int_0^\infty k_t(x - y)\, dt\right) f(y)\, dy.
\end{aligned}
$$

DOI: 10.1201/9781003206781-8

For $x \in \mathbb{R}^n \setminus \{0\}$,

$$
\begin{aligned}
\int_0^\infty k_t(x)\, dt &= \int_0^\infty (4\pi t)^{-n/2} e^{-|x|^2/(4t)}\, dt \\
&= (4\pi)^{-n/2} \int_0^\infty e^{-s|x|^2/4} s^{(n/2)-1} \frac{ds}{s} \\
&= \frac{|x|^{2-n}}{4\pi^{n/2}} \int_0^\infty e^{-t} t^{(n/2)-1} \frac{dt}{t} = \frac{\Gamma\left(\frac{n-2}{2}\right)}{4\pi^{n/2}} |x|^{2-n}
\end{aligned}
$$

provided that $n \geq 3$. Thus,

$$
(\Delta^{-1} f)(x) = \int_{\mathbb{R}^n} N_n(x-y) f(y)\, dy, \quad x \in \mathbb{R}^n,
$$

where

$$
N_n(x) = -\frac{\Gamma\left(\frac{n-2}{2}\right)}{4\pi^{n/2}} |x|^{2-n}, \quad x \in \mathbb{R}^n \setminus \{0\},
$$

for $n \geq 3$. To understand the constant in front of $|x|^{2-n}$, let us use the recurrence formula in Theorem 2.3 to obtain

$$
\Gamma\left(\frac{n-2}{2}\right) = \Gamma\left(\frac{n}{2} - 1\right) = \frac{\Gamma\left(\frac{n}{2}\right)}{\frac{n}{2} - 1} = \frac{2\Gamma\left(\frac{n}{2}\right)}{n-2}.
$$

Thus, using the formula for the surface area $|\mathbb{S}^{n-1}|$ of the unit sphere $\mathbb{S}^{n-1}$ in $\mathbb{R}^n$ given in Theorem 2.8, we see that for all x in $\mathbb{R}^n \setminus \{0\}$,

$$
N_n(x) = -\frac{\Gamma\left(\frac{n}{2}\right)}{(n-2)2\pi^{n/2}} |x|^{2-n} = \frac{1}{(2-n)|\mathbb{S}^{n-1}|} |x|^{2-n}.
$$

In order to obtain a formula for $\Delta^{-1} f$ when $n = 2$, we let

$$
\omega(x) = \int_0^\infty (4\pi t)^{-1} e^{-|x|^2/(4t)}\, dt, \quad x \in \mathbb{R}^2.
$$

Let $r = |x|$ and we are led to look at the function ω on $(0, \infty)$ defined by

$$
\omega(r) = \frac{1}{4\pi} \int_0^\infty \frac{1}{t} e^{-r^2/(4t)}\, dt, \quad r \in (0, \infty).
$$

Note that

$$
\begin{aligned}
\omega'(r) &= -\frac{1}{8\pi} \int_0^\infty \frac{1}{t} e^{-r^2/(4t)} r \frac{dt}{t} \\
&= -\frac{1}{8\pi} \int_0^\infty e^{-sr^2/4} r\, ds = \frac{1}{2\pi r} e^{-sr^2/4} \Big|_0^\infty = -\frac{1}{2\pi r}
\end{aligned}
$$

for all r in $(0, \infty)$. Therefore

$$
\omega(r) = -\frac{1}{2\pi} \ln r + C, \quad r \in (0, \infty),
$$

where C is an arbitrary constant. If we choose $C = 0$, then

$$(\Delta^{-1}f)(x) = \int_{\mathbb{R}^2} N_2(x-y)f(y)\,dy, \quad x \in \mathbb{R}^2,$$

where

$$N_2(x) = \frac{1}{2\pi}\ln|x| = \frac{1}{|\mathbb{S}^1|}\ln|x|, \quad x \in \mathbb{R}^2 \setminus \{0\}.$$

The function N_n is known as the Newtonian potential or the Green function for the Laplacian Δ on $\mathbb{R}^n$. It is the negative of the integral of the heat kernel from 0 to ∞ with respect to time. In cosmological terms, it is the gravitational potential generated by a heavy mass placed at the center.

The derivation of the formula for a solution of the Poisson equation in (8.1) is by no means rigorous, but is highly suggestive that the resulting formula is indeed a solution. We are then led to the following theorem.

Theorem 8.1 *Let $f \in \mathcal{S}$. Then the function u on $\mathbb{R}^n$ defined by*

$$u = N_n * f$$

is a solution of the Poisson equation (8.1).

Proof By Exercise 1 in this chapter, N_n is a tempered distribution. So, by Remark 5.11, $u = N_n * f$ is a tempered function on $\mathbb{R}^n$. By Exercises 15 and 16 in Chapter 5,

$$\widehat{\Delta u} = (\Delta(N_n * f))^\wedge = -|\cdot|^2(N_n * f)^\wedge = -(2\pi)^{n/2}|\cdot|^2\hat{f}\widehat{N_n}. \tag{8.2}$$

Now, for all φ in $\mathcal{S}$, we get, by the adjoint formula for the Fourier transform,

$$\begin{aligned}
\widehat{N_n}(\varphi) &= N_n(\hat{\varphi}) = \int_{\mathbb{R}^n} N_n(\xi)\hat{\varphi}(\xi)\,d\xi \\
&= -\int_{\mathbb{R}^n}\left(\int_0^\infty k_t(\xi)\,dt\right)\hat{\varphi}(\xi)\,d\xi \\
&= -\int_0^\infty\left(\int_{\mathbb{R}^n} k_t(\xi)\hat{\varphi}(\xi)\,d\xi\right)dt \\
&= -\int_0^\infty\left(\int_{\mathbb{R}^n} \widehat{k_t}(\xi)\,\varphi(\xi)\,d\xi\right)dt.
\end{aligned}$$

So, for all φ in $\mathcal{S}$,

$$\widehat{N_n}(\varphi) = -(2\pi)^{-n/2}\int_0^\infty\left(\int_{\mathbb{R}^n} e^{-t|\eta|^2}\varphi(\eta)\,d\eta\right)dt. \tag{8.3}$$

But, by (8.3),

$$
\begin{aligned}
(|\cdot|^2 \widehat{N_n})(\varphi) &= \widehat{N_n}(|\cdot|^2 \varphi) \\
&= -(2\pi)^{-n/2} \int_0^\infty \left(\int_{\mathbb{R}^n} e^{-t|\xi|^2} |\xi|^2 \varphi(\xi)\, d\xi \right) dt \\
&= -(2\pi)^{-n/2} \int_0^\infty \left(-\frac{d}{dt} \int_{\mathbb{R}^n} e^{-t|\xi|^2} \varphi(\xi)\, d\xi \right) dt \\
&= (2\pi)^{-n/2} \int_{\mathbb{R}^n} e^{-t|\xi|^2} \varphi(\xi)\, d\xi \Big|_{t=0}^\infty \\
&= -(2\pi)^{-n/2} \int_{\mathbb{R}^n} \varphi(\xi)\, d\xi.
\end{aligned}
$$

This gives

$$
|\cdot|^2 \widehat{N_n} = -(2\pi)^{-n/2}.
$$

By (8.2), we get

$$
\widehat{\Delta u} = (\Delta(N_n * f))^\wedge = -(2\pi)^{n/2} |\cdot|^2 \hat{f} \widehat{N_n} = \hat{f}.
$$

By the Fourier inversion formula, we get

$$
\Delta u = f
$$

and the proof is complete. $\square$

Historical Notes

Siméon-Denis Poisson (1781–1840), a French mathematician and physicist, was a student of Laplace at École Polytechnique. His scientific life and political ideas were very similar to those of Laplace and he was well liked by Laplace. Once he graduated from École Polytechnique in 1800, he was offered a teaching position there. He replaced Fourier there in 1806. His scientific contributions include the works of Laplace and others on celestial mechanics and the work of Fourier on heat. He is well known for the Poisson distribution in probability and the Poisson kernel in analysis, among others.

Isaac Newton (1642–1727), a household name in mathematics and physics, was educated at Trinity College of the University of Cambridge and appointed Lucasian Professor of Mathematics at the University of Cambridge in 1669. He is best known for his creation of calculus and discovery of the law of gravitation. His most famous book *Philosophiae Naturalis Principia Mathematica* is still widely read by scholars in mathematics, science, and philosophy. The impact of his scientific ideas on modern civilization is profound.

George Green (1793–1841), a British mathematical physicist, was born in Nottingham, England. He was self-taught and became known because of *An Essay on the Application of Mathematical Analysis to the Theories of*

Electricity and Magnetism published by him in 1828. He is best known today for Green's theorem in vector integral calculus and Green functions in partial differential equations and mathematical physics.

Exercises

1. Prove that for every complex number α, the function $\tilde{N}_n$ on $\mathbb{R}^n$ defined by

$$\tilde{N}_n(x) = \begin{cases} N_n(x), & x \neq 0, \\ \alpha, & x = 0, \end{cases}$$

is a tempered function on $\mathbb{R}^n$.

2. Compute ΔN_n. (Hint: Use the fact that $u = N_n * f$ is a solution of

$$\Delta u = f$$

on $\mathbb{R}^n$ for all functions f in $\mathcal{S}$.)

3. Find a solution of the partial differential equation

$$\frac{\partial u}{\partial x_1} + i\frac{\partial u}{\partial x_2} = f$$

on $\mathbb{R}^2$ in terms of the Newtonian potential N_2, where f is a given Schwartz function on $\mathbb{R}^2$.

4. Let $f_1, f_2, \ldots, f_n$ be Schwartz functions on $\mathbb{R}^n$ such that the vector field $\mathbf{f} = (f_1, f_2, \ldots, f_n)$ on $\mathbb{R}^n$ is conservative. Use the Newtonian potential to find a solution u of the partial differential equation

$$\nabla u = \mathbf{f}$$

on $\mathbb{R}^n$.

5. Let f be a Schwartz function on $\mathbb{R}^n$. Use the Newtonian potential to find a solution u in $C^\infty(\mathbb{R}^n)$ of the fourth-order partial differential equation

$$\Delta^2 u = f$$

on $\mathbb{R}^n$.

Chapter 9

The Bessel Potential

From Chapter 8, we know that Δ is a negative operator. In applications in quantum mechanics and differential geometry, it is more customary to work with the positive operator $-\Delta$ instead and we follow this practice.

We construct in this chapter in two different ways the kernel of the inverse of the perturbed Laplacian $-\Delta + m^2$, where m is a positive constant. The answer comes from the family of functions known as the Bessel potentials.

Theorem 9.1 Let $s > 0$. If we define the function $G_{s,m}$ on $\mathbb{R}^n$ by

$$G_{s,m}(x) = \frac{m^{n-s}}{2^{s/2}\Gamma(s/2)} \int_0^\infty e^{-r/2} e^{-m^2|x|^2/(2r)} r^{-(n-s)/2} \frac{dr}{r}, \quad x \in \mathbb{R}^n,$$

then

$$G_{s,m} \in L^1(\mathbb{R}^n),$$

$$\|G_{s,m}\|_1 = (2\pi)^{n/2} m^{-s},$$

and

$$\widehat{G_{s,m}}(\xi) = (m^2 + |\xi|^2)^{-s/2}, \quad \xi \in \mathbb{R}^n.$$

Proof Since

$$G_{s,m}(x) \geq 0, \quad x \in \mathbb{R}^n,$$

it follows from Fubini's theorem that

$$
\begin{aligned}
\|G_{s,m}\|_1 &= \frac{m^{n-s}}{2^{s/2}\Gamma(s/2)} \int_{\mathbb{R}^n} \left(\int_0^\infty e^{-r/2} e^{-m^2|x|^2/(2r)} r^{-(n-s)/2} \frac{dr}{r} \right) dx \\
&= \frac{m^{n-s}}{2^{s/2}\Gamma(s/2)} \int_0^\infty e^{-r/2} r^{-(n-s)/2} \left(\int_{\mathbb{R}^n} e^{-m^2|x|^2/(2r)} dx \right) \frac{dr}{r}.
\end{aligned}
$$

Since

$$\int_{\mathbb{R}^n} e^{-m^2|x|^2/(2r)} dx = (2\pi r)^{n/2} m^{-n},$$

we get

$$
\begin{aligned}
\|G_{s,m}\|_1 &= \frac{(2\pi)^{n/2} m^{-s}}{2^{s/2}\Gamma(s/2)} \int_0^\infty e^{-r/2} r^{s/2} \frac{dr}{r} \\
&= \frac{(2\pi)^{n/2} m^{-s}}{2^{s/2}\Gamma(s/2)} 2^{s/2} \int_0^\infty e^{-r} r^{s/2} \frac{dr}{r} = (2\pi)^{n/2} m^{-s}.
\end{aligned}
$$

DOI: 10.1201/9781003206781-9

Using the definition of the Fourier transform, we get

$$\widehat{G_{s,m}}(\xi)$$
$$= (2\pi)^{-n/2} \int_{\mathbb{R}^n} e^{-ix\cdot\xi} G_s(x)\, dx$$
$$= \frac{(2\pi)^{-n/2} m^{n-s}}{2^{s/2}\Gamma(s/2)} \int_0^\infty e^{-r/2} \left(\int_{\mathbb{R}^n} e^{-ix\cdot\xi} e^{-m^2|x|^2/(2r)}\, dx \right) r^{-(n-s)/2} \frac{dr}{r}$$
$$= \frac{m^{n-s}}{2^{s/2}\Gamma(s/2)} \int_0^\infty e^{-r/2} r^{n/2} e^{-r|\xi|^2/(2m^2)} r^{-(n-s)/2} \frac{dr}{r}$$
$$= \frac{m^{-s}}{2^{s/2}\Gamma(s/2)} \int_0^\infty e^{-\frac{r}{2}\left(1 + \frac{|\xi|^2}{m^2}\right)} r^{s/2} \frac{dr}{r}$$
$$= \frac{m^{-s}}{2^{s/2}\Gamma(s/2)} \int_0^\infty e^{-r} (2r)^{s/2} \frac{dr}{r} \left(1 + \frac{|\xi|^2}{m^2}\right)^{-s/2} = (m^2 + |\xi|^2)^{-s/2}$$

for all ξ in $\mathbb{R}^n$. In the above computations, we use the third formula in Proposition 4.5 and Proposition 4.6 on the Fourier transform of the function $e^{-|x|^2/2}$, $x \in \mathbb{R}^n$, to obtain the fact that

$$(2\pi)^{-n/2} \int_{\mathbb{R}^n} e^{-ix\cdot\xi} e^{-m^2|x|^2/(2r)}\, dx = m^{-n} r^{n/2} e^{-r|\xi|^2/(2m^2)}, \quad \xi \in \mathbb{R}^n.$$

$\square$

We can now find a solution of the partial differential equation

$$(-\Delta + m^2)u = f$$

on $\mathbb{R}^n$, where m is a positive constant and $f \in \mathcal{S}$. Taking the Fourier transform on both sides of the partial differential equation gives

$$(m^2 + |\xi|^2)\hat{u}(\xi) = \hat{f}(\xi), \quad \xi \in \mathbb{R}^n.$$

Therefore

$$\hat{u}(\xi) = \frac{\hat{f}(\xi)}{m^2 + |\xi|^2}, \quad \xi \in \mathbb{R}^n.$$

Taking the inverse Fourier transform, and using Proposition 4.1 and Theorem 9.1, we get

$$u(x) = (2\pi)^{-n/2} (G_{2,m} * f)(x), \quad x \in \mathbb{R}^n.$$

In other words, the kernel of the inverse of $-\Delta + m^2$ is $(2\pi)^{-n/2} G_{2,m}$.

Remark 9.2 If $m = 1$, then we are back to the Bessel potential $G_{s,1}$ studied in [36, 38, 58]. The attribution to Bessel is due to the fact that these functions can be represented in terms of Bessel functions. We find it more convenient to represent these functions in terms of the integral given in Theorem 9.1.

We can get the same kernel by the same method used in constructing the Newtonian potential. In other words, we can get the kernel of the inverse of $-\Delta + m^2$ by integrating from 0 to ∞ with respect to time of the heat kernel for the operator $\Delta - m^2$. Indeed, as in the case of the heat kernel for the Laplacian, the heat kernel $h_t(x)$, $x \in \mathbb{R}^n$, $t > 0$, of the operator $\Delta - m^2$ is given by

$$h_t(x) = e^{-m^2 t} k_t(x) = e^{-m^2 t} (4\pi t)^{-n/2} e^{-|x|^2/4t}, \quad x \in \mathbb{R}^n, t > 0.$$

Therefore the kernel h of the inverse of $-\Delta + m^2$ is given by

$$
\begin{aligned}
h(x) &= \int_0^\infty h_t(x)\, dt \\
&= \int_0^\infty e^{-m^2 t} (4\pi t)^{-n/2} e^{-|x|^2/(4t)}\, dt \\
&= (4\pi)^{-n/2} \int_0^\infty e^{-m^2 t} t^{-(n-2)/2} e^{-|x|^2/(4t)} \frac{dt}{t}
\end{aligned}
$$

for all x in $\mathbb{R}^n$. If we let $m^2 t = \frac{r}{2}$, then

$$
\begin{aligned}
h(x) &= (4\pi)^{-n/2} \int_0^\infty e^{-r/2} \left(\frac{r}{2m^2}\right)^{-(n-2)/2} e^{-m^2|x|^2/2r} \frac{dr}{r} \\
&= (2\pi)^{-n/2} \frac{m^{n-2}}{2} \int_0^\infty e^{-r/2} e^{-m^2|x|^2/2r} r^{-(n-2)/2} \frac{dr}{r} \\
&= (2\pi)^{-n/2} G_{2,m}(x), \quad x \in \mathbb{R}^n.
\end{aligned}
$$

The existence and properties of solutions of the partial differential equation

$$(-\Delta + m^2) u = f$$

on $\mathbb{R}^n$, where $f \in L^p(\mathbb{R}^n)$, $1 \le p \le \infty$, are given in Exercises 1 and 2 in this chapter.

Historical Notes

Friedrich Wilhelm Bessel (1784–1846) was a German mathematician. He is best known for one of the most important special functions in mathematics, which is known as the Bessel function. He also made contributions to astronomy and was appointed Director of the Observatory in Köningsberg in 1810.

Exercises

1. Let $f \in L^p(\mathbb{R}^n)$, $1 \le p \le \infty$. Prove that the function u on $\mathbb{R}^n$ given by

$$u = (2\pi)^{-n/2}(G_{2,m} * f)$$

   is a solution of the partial differential equation

$$(-\Delta + m^2) u = f$$

   on $\mathbb{R}^n$.

2. Let u and f be as in Exercise 1. Prove that

$$\|u\|_p \le m^{-2}\|f\|_p, \quad 1 \le p \le \infty.$$

3. Find the kernel of the inverse of $(-\Delta + m^2)^k$, where m is a positive constant and k is a positive integer.

Chapter 10

Global Hypoellipticity in the Schwartz Space

We give in this chapter two proofs of Theorem 10.1 on the global hypoellipticity in $\mathcal{S}$ of the perturbed Laplacian $-\Delta + m^2$, where m is a positive constant. The first proof is elementary and the second proof is deeper in the sense that it probes the behavior of the Bessel potential at infinity. Both proofs depend on Lemma 10.3.

Theorem 10.1 *Let m be a positive number. Then $-\Delta + m^2$ is globally hypolliptic in the sense that*

$$u \in \mathcal{S}', \ (-\Delta + m^2)u \in \mathcal{S} \Rightarrow u \in \mathcal{S}.$$

Remark 10.2 The global hypoellipticity in Theorem 10.1 means that every tempered distribution u for which $(-\Delta + m^2)u$ is a Schwartz function on $\mathbb{R}^n$ must be a Schwartz function on $\mathbb{R}^n$. So, every solution u in $\mathcal{S}'$ of the partial differential equation

$$(-\Delta + m^2)u = f, \quad f \in \mathcal{S},$$

on $\mathbb{R}^n$ has the same smoothness and global behavior as the given function f. "Hypoellipticity" is the word coined for the property of $-\Delta + m^2$ to the effect that

$$f \in C^\infty(\mathbb{R}^n) \Rightarrow u \in C^\infty(\mathbb{R}^n).$$

More details on hypoellipticity are given in Chapter 18. The global hypoellipticity of another important operator in quantum physics is given in Chapter 16. Results on global hypoellipticity in $\mathcal{S}$ for very general operators can be found in [39].

Theorem 10.1 is obviously false for the Laplacian $-\Delta$. This can be seen by observing that the function u on $\mathbb{R}^n$ given by

$$u(x) = 1, \quad x \in \mathbb{R}^n,$$

is such that $\Delta u = 0$. Certainly, $u \notin \mathcal{S}$.

DOI: 10.1201/9781003206781-10

Lemma 10.3 *Let $\varphi \in \mathcal{S}$. Then the function ψ defined on $\mathbb{R}^n$ by*

$$\psi(\xi) = \frac{\varphi(\xi)}{m^2 + |\xi|^2}, \quad \xi \in \mathbb{R}^n,$$

is also in $\mathcal{S}$.

The following formula for the derivative of the reciprocal of a function, to be used in the proof of Lemma 10.3, is a useful formula in its own right.

Lemma 10.4 *Let $f \in C^\infty(\mathbb{R}^n)$. Then for all multi-indices α,*

$$\partial^\alpha \left(\frac{1}{f} \right) = \sum C_{\alpha^{(1)}, \alpha^{(2)}, \ldots, \alpha^{(k)}} \frac{(\partial^{\alpha^{(1)}} f)(\partial^{\alpha^{(2)}} f) \cdots (\partial^{\alpha^{(k)}} f)}{f^{k+1}},$$

where $C_{\alpha^{(1)}, \alpha^{(2)}, \ldots, \alpha^{(k)}}$ is a constant depending only on $\alpha^{(1)}, \alpha^{(2)}, \ldots, \alpha^{(k)}$ and the sum is taken over all multi-indices $\alpha^{(1)}, \alpha^{(2)}, \ldots, \alpha^{(k)}$, which form a partition of α.

Proof The formula is obviously true for the zero multi-index. Suppose that the formula is true for all multi-indices α with $|\alpha| = m$. Let β be a multi-index such that $|\beta| = m + 1$. Then we can write

$$\partial^\beta = \partial_j \partial^\alpha,$$

where α is a multi-index with $|\alpha| = m$ and $1 \leq j \leq n$. By the induction hypothesis,

$$\partial^\alpha \left(\frac{1}{f} \right) = \sum C_{\alpha^{(1)}, \ldots, \alpha^{(k)}} \frac{(\partial^{\alpha^{(1)}} f) \cdots (\partial^{\alpha^{(k)}} f)}{f^{k+1}} \qquad (10.1)$$

as claimed in the lemma. Let $\mathbf{j}$ be the multi-index such that the jth entry is 1 and the other entries are equal to zero. We first observe that every partition of β is of the form

$$\alpha^{(1)} + \alpha^{(2)} + \cdots + \alpha^{(k)} + \mathbf{j}$$

or

$$\gamma^{(1)} + \alpha^{(2)} + \cdots + \alpha^{(k)}$$

or

$$\alpha^{(1)} + \gamma^{(2)} + \cdots + \alpha^{(k)}$$

or

$$\cdots$$

or

$$\alpha^{(1)} + \alpha^{(2)} + \cdots + \gamma^{(k)},$$

where

$$\gamma^{(l)} = \alpha^{(l)} + \mathbf{j}$$

and $\alpha^{(1)}, \alpha^{(2)}, \ldots, \alpha^{(k)}$ form a partition of α. Using (10.1), we obtain

$$
\begin{aligned}
\partial^{\beta}\left(\frac{1}{f}\right) &= \sum C_{\alpha^{(1)},\ldots,\alpha^{(k)}}(k+1)\frac{(\partial^{\alpha^{(1)}}f)\cdots(\partial^{\alpha^{(k)}}f)f^{k}(\partial_{j}f)}{f^{2(k+1)}} + \\
&\quad \sum C_{\alpha^{(1)},\ldots,\alpha^{(k)}}\frac{f^{k+1}\sum_{l=1}^{k}(\partial^{\alpha^{(1)}}f)\cdots(\partial^{\gamma^{(l)}}f)\cdots(\partial^{\alpha^{(k)}}f)}{f^{2(k+1)}} \\
&= \sum C_{\alpha^{(1)},\ldots,\alpha^{(k)}}(k+1)\frac{(\partial^{\alpha^{(1)}}f)\cdots(\partial^{\alpha^{(k)}}f)(\partial^{j}f)}{f^{(k+1)+1}} + \\
&\quad \sum C_{\alpha^{(1)},\ldots,\alpha^{(k)}}\frac{\sum_{l=1}^{k}(\partial^{\alpha^{(1)}}f)\cdots(\partial^{\gamma^{(l)}}f)\cdots(\partial^{\alpha^{(k)}}f)}{f^{k+1}},
\end{aligned}
$$

which is the same as

$$
\partial^{\beta}\left(\frac{1}{f}\right) = \sum C_{\beta^{(1)},\ldots,\beta^{(k)}}\frac{(\partial^{\beta^{(1)}}f)\cdots(\partial^{\beta^{(k)}}f)}{f^{k+1}},
$$

where $C_{\beta^{(1)},\ldots,\beta^{(k)}}$ is a constant depending only on $\beta^{(1)},\ldots,\beta^{(k)}$ and the sum is taken over all $\beta^{(1)},\ldots,\beta^{(k)}$ partitioning β. $\qquad\square$

Proof of Lemma 10.3 Let α and β be given multi-indices. Let σ be the function on $\mathbb{R}^{n}$ given by

$$
\sigma(\xi) = m^{2} + |\xi|^{2}, \quad \xi \in \mathbb{R}^{n}.
$$

Then, using the Leibniz formula given in Proposition 1.1 and Lemma 10.4, we get

$$
(\xi^{\alpha}\partial^{\beta}\psi)(\xi) = \xi^{\alpha}\sum_{\gamma \le \beta}\binom{\beta}{\gamma}(\partial^{\beta-\gamma}\varphi)(\xi)\sum C_{\gamma^{(1)},\ldots,\gamma^{(k)}}\frac{(\partial^{\gamma^{(1)}}\sigma)(\xi)\cdots(\partial^{\gamma^{(k)}}\sigma)(\xi)}{(\sigma(\xi))^{k+1}}
$$

for all ξ in $\mathbb{R}^{n}$, where $C_{\gamma^{(1)},\ldots,\gamma^{(k)}}$ is a constant depending only on the multi-indices $\gamma^{(1)},\ldots,\gamma^{(k)}$ and the summation $\sum$ is taken over all multi-indices $\gamma^{(1)},\ldots,\gamma^{(k)}$, which partition γ. Thus,

$$
\begin{aligned}
&|\xi^{\alpha}(\partial^{\beta}\psi)(\xi)| \\
&\le \sum_{\gamma \le \beta}\sum\binom{\beta}{\gamma}|C_{\gamma^{(1)},\ldots,\gamma^{(k)}}||\xi^{\alpha}(\partial^{\beta-\gamma}\varphi)(\xi)|\frac{|(\partial^{\gamma^{(1)}}\sigma)(\xi)\cdots(\partial^{\gamma^{(k)}}\sigma)(\xi)|}{m^{2(k+1)}}
\end{aligned}
$$

for all ξ in $\mathbb{R}^{n}$. Since $\partial^{\gamma^{(1)}}\sigma,\ldots,\partial^{\gamma^{(k)}}\sigma$ are polynomials and $\varphi \in \mathcal{S}$, it follows that

$$
\sup_{\xi \in \mathbb{R}^{n}}|\xi^{\alpha}(\partial^{\beta}\psi)(\xi)| < \infty.
$$

$\qquad\square$

First Proof of Theorem 10.1 Let $u \in \mathcal{S}'$ be such that $(-\Delta + m^2)u \in \mathcal{S}$. Let

$$f = (-\Delta + m^2)u.$$

Then, using the Fourier transform,

$$(m^2 + |\cdot|^2)\hat{u} = \hat{f}.$$

So,

$$\hat{u} = \frac{\hat{f}}{m^2 + |\cdot|^2}.$$

Since $\hat{f} \in \mathcal{S}$, we can use Lemma 10.3 to conclude that $\hat{u} \in \mathcal{S}$. So, by the Fourier inversion formula, $u \in \mathcal{S}$. □

Now, we give a pointwise estimate for the Bessel potential $G_{s,m}$, $s > 0$, outside the unit ball with center at the origin.

Theorem 10.5 *Let m and s be positive numbers. Then for every positive number a in $(0, m)$, there exists a positive constant $C_{s,m,a}$ such that*

$$G_{s,m}(x) \leq C_{s,m,a} e^{-a|x|}, \quad |x| > 1.$$

Proof If we let $t = r/2$ in the definition of $G_{s,m}$, then we get

$$G_{s,m}(x) = \frac{m^{n-s}}{2^{n/2}\Gamma(s/2)} \int_0^\infty e^{-t} e^{-m^2|x|^2/(4t)} t^{-(n-s)/2} \frac{dt}{t}, \quad x \in \mathbb{R}^n.$$

If we let $\rho = 2t/|x|$ in the preceding equation, then we get

$$
\begin{aligned}
G_{s,m}(x) &= \frac{m^{n-s}}{2^{n/2}\Gamma(s/2)} \int_0^\infty e^{-\rho|x|/2} e^{-m^2|x|/(2\rho)} (\rho|x|/2)^{-(n-s)/2} \frac{d\rho}{\rho} \\
&= \frac{m^{n-s}}{2^{s/2}\Gamma(s/2)} |x|^{(s-n)/2} \int_0^\infty e^{-\rho|x|/2} e^{-m^2|x|/(2\rho)} \rho^{(s-n)/2} \frac{d\rho}{\rho} \\
&= \frac{m^{n-s}}{2^{s/2}\Gamma(s/2)} |x|^{(s-n)/2} \int_0^\infty e^{-|x|(\rho^2+m^2)/(2\rho)} \rho^{(s-n)/2} \frac{d\rho}{\rho} \\
&= \frac{m^{n-s}}{2^{s/2}\Gamma(s/2)} |x|^{(s-n)/2} e^{-m|x|} \int_0^\infty e^{-|x|(\rho-m)^2/(2\rho)} \rho^{(s-n)/2} \frac{d\rho}{\rho}
\end{aligned}
$$

for all x in $\mathbb{R}^n \setminus \{0\}$. So, for $|x| \geq 1$,

$$G_{s,m}(x) \leq \frac{m^{n-s}}{2^{s/2}\Gamma(s/2)} |x|^{(s-n)/2} e^{-m|x|} \int_0^\infty e^{-(\rho-m)^2/(2\rho)} \rho^{(s-n)/2} \frac{d\rho}{\rho}.$$

Therefore for all $a \in (0, m)$, we can find a positive constant $C_{s,m,a}$ such that

$$G_{s,m}(x) \leq C_{s,m,a} e^{(m-a)|x|} e^{-m|x|} = C_{s,m,a} e^{-a|x|}, \quad |x| \geq 1,$$

and the proof is complete. □

Second Proof of Theorem 10.1 First we note that if u_1 and u_2 are tempered distributions such that

$$(-\Delta + m^2)u_1 = (-\Delta + m^2)u_2,$$

then

$$u_1 = u_2.$$

Indeed, let $u = u_1 - u_2$. Then

$$(-\Delta + m^2)u = 0.$$

Applying the Fourier transform to both sides of the equation, we get

$$(m^2 + |\cdot|^2)\hat{u} = 0.$$

For all φ in $\mathcal{S}$, Lemma 10.3 implies that the function $(m^2 + |\cdot|^2)^{-1}\varphi$ is also in $\mathcal{S}$. So,

$$(m^2 + |\cdot|^2)\hat{u}((m^2 + |\cdot|^2)^{-1}\varphi) = 0$$

and hence

$$\hat{u}(\varphi) = 0.$$

Therefore $\hat{u} = 0$. Using the Fourier inversion formula for tempered distributions, we get $u = 0$. Thus, $u_1 = u_2$. Now, let $u \in \mathcal{S}'$ be such that $(-\Delta + m^2)u \in \mathcal{S}$. Let $f = (-\Delta + m^2)u$. Then, by the uniqueness of solutions, we get

$$u(x) = (2\pi)^{-n/2}(G_{2,m} * f)(x) = (2\pi)^{-n/2}\int_{\mathbb{R}^n} G_{2,m}(y)f(x-y)\,dy, \quad x \in \mathbb{R}^n.$$

Let β be any multi-index. Then for all x in $\mathbb{R}^n$, we get, by Young's inequality and the fact that $\|G_{2,m}\|_1 = (2\pi)^{n/2}m^{-2}$,

$$
\begin{aligned}
|(\partial^\beta u)(x)| &\leq (2\pi)^{-n/2}\int_{\mathbb{R}^n} |G_{2,m}(y)(\partial^\beta f)(x-y)|\,dy \\
&\leq (2\pi)^{-n/2}\sup_{y\in\mathbb{R}^n} |(\partial^\beta f)(y)|\,\|G_{2,m}\|_1 \\
&= m^{-2}\sup_{y\in\mathbb{R}^n} |(\partial^\beta f)(y)|.
\end{aligned}
$$

Thus,

$$\sup_{x\in\mathbb{R}^n} |(\partial^\beta u)(x)| < \infty.$$

Let α and β be arbitrary multi-indices with $\alpha \neq 0$. Then for all x in $\mathbb{R}^n$, we use the simple inequality from the first exercise in Chapter 1, *i.e.*,

$$|x^\alpha| \leq |x|^{|\alpha|},$$

to get

$$
\begin{aligned}
|x^\alpha(\partial^\beta u)(x)| &\leq (2\pi)^{-n/2} \int_{\mathbb{R}^n} |x^\alpha| \, |G_{2,m}(y)| \, |(\partial^\beta f)(x-y)| \, dy \\
&= (2\pi)^{-n/2} \int_{\mathbb{R}^n} |x|^{|\alpha|} |G_{2,m}(y)| \, |(\partial^\beta f)(x-y)| \, dy \\
&= (2\pi)^{-n/2} \int_{\mathbb{R}^n} |(x-y)+y|^{|\alpha|} |G_{2,m}(y)| \, |(\partial^\beta f)(x-y)| \, dy \\
&\leq (2\pi)^{-n/2} 2^{|\alpha|} (I(x)+J(x)),
\end{aligned}
$$

where

$$
I(x) = \int_{\mathbb{R}^n} |G_{2,m}(y)| \, |x-y|^{|\alpha|} |(\partial^\beta f)(x-y)| \, dy
$$

and

$$
J(x) = \int_{\mathbb{R}^n} |y|^{|\alpha|} |G_{2,m}(y)| \, |(\partial^\beta f)(x-y)| \, dy.
$$

Since $\|G_{2,m}\|_1 = (2\pi)^{n/2} m^{-2}$, we see that

$$
I(x) \leq (2\pi)^{n/2} m^{-2} \sup_{y \in \mathbb{R}^n} \{|y|^{|\alpha|} |(\partial^\beta f)(y)|\}, \quad x \in \mathbb{R}^n.
$$

So,

$$
\sup_{x \in \mathbb{R}^n} I(x) < \infty.
$$

Moreover,

$$
J(x) \leq \sup_{y \in \mathbb{R}^n} |(\partial^\beta f)(y)| \int_{\mathbb{R}^n} |y|^{|\alpha|} |G_{2,m}(y)| \, dy, \quad x \in \mathbb{R}^n.
$$

Write

$$
\int_{\mathbb{R}^n} |y|^{|\alpha|} |G_{2,m}(y)| \, dy = \left(\int_{|y| \leq 1} + \int_{|y| \geq 1} \right) |y|^{|\alpha|} |G_{2,m}(y)| \, dy.
$$

Since $\|G_{2,m}\|_1 = (2\pi)^{n/2} m^{-2}$,

$$
\int_{|y| \leq 1} |y|^{|\alpha|} |G_{2,m}(y)| \, dy \leq \int_{|y| \leq 1} |G_{2,m}(y)| \, dy \leq \|G_{2,m}\|_1 = (2\pi)^{n/2} m^{-2}.
$$

Let $a \in (0, m)$. Then, by Theorem 10.5, there exists a positive constant $C_{m,a}$ such that

$$
\int_{|y| \geq 1} |y|^{|\alpha|} |G_{2,m}(y)| \, dy \leq C_{m,a} \int_{|y| \geq 1} |y|^{|\alpha|} e^{-a|y|} dy < \infty.
$$

So,

$$
\sup_{x \in \mathbb{R}^n} J(x) \leq \sup_{y \in \mathbb{R}^n} |(\partial^\beta f)(y)| \left(\int_{|y| \leq 1} + \int_{|y| \geq 1} \right) |y|^{|\alpha|} |G_{2,m}(y)| \, dy < \infty.
$$

Hence
$$\sup_{x\in\mathbb{R}^n} |x^\alpha(\partial^\beta u)(x)| < \infty$$
and this completes the proof that $u \in \mathcal{S}$. $\qquad\qquad\square$

Exercises

1. Is the Laplacian Δ globally hypoelliptic on $\mathbb{R}^n$?

2. Let m be a positive constant and let k be a positive integer. Prove that the partial differential operator $(-\Delta + m^2)^k$ is globally hypoelliptic on $\mathbb{R}^n$.

3. Let m be a positive constant. Is the partial differential operator $-\Delta - m^2$ globally hypoelliptic on $\mathbb{R}^n$? (Hint: You may want to pick a point ξ in the sphere $\{\xi \in \mathbb{R}^n : |\xi| = m\}$ and construct a solution of the partial differential equation
$$(-\Delta - m^2)u = 0$$
on $\mathbb{R}^n$.)

Chapter 11

The Poisson Kernel

In this chapter we seek a solution $u = u(x, t)$, $x \in \mathbb{R}^n, t > 0$, of the Dirichlet problem for the Laplacian on the upper half space $\{(x, t) : x \in \mathbb{R}^n, t > 0\}$ given by

$$\begin{cases} \frac{\partial^2 u}{\partial t^2}(x, t) + (\Delta u)(x, t) = 0, & x \in \mathbb{R}^n, \, t > 0, \\ u(x, 0) = f(x), & x \in \mathbb{R}^n, \end{cases} \tag{11.1}$$

where $f \in \mathcal{S}$. The problem in (11.1) is different in nature from the ones in Chapters 6 and 7. The $t = 0$ in Chapters 6 and 7 refers to the initial time and $t = 0$ in this chapter is understood to be the boundary of the upper half space $\{(x, t) : x \in \mathbb{R}^n, \, t > 0\}$. The problem in this chapter is an example of a boundary value problem.

Taking the partial Fourier transform with respect to x, we get

$$\frac{\partial^2 \hat{u}}{\partial t^2}(\xi, t) - |\xi|^2 \hat{u}(\xi, t) = 0, \quad \xi \in \mathbb{R}^n, \, t > 0,$$

and hence

$$\hat{u}(\xi, t) = C_1 e^{t|\xi|} + C_2 e^{-t|\xi|}, \quad \xi \in \mathbb{R}^n, \, t > 0.$$

For $t > 0$, the function $e^{t|\xi|}$, $\xi \in \mathbb{R}^n$, is a not a tempered function on $\mathbb{R}^n$ and hence is beyond the scope of the applicability of the Fourier transform. So, we let $C_1 = 0$ and look at

$$\hat{u}(\xi, t) = C_2 e^{-t|\xi|}, \quad \xi \in \mathbb{R}^n, \, t > 0.$$

Using the boundary condition of the Dirichlet problem, we get

$$C_2 = \hat{f}(\xi)$$

and therefore

$$\hat{u}(\xi, t) = \hat{f}(\xi) e^{-t|\xi|}, \quad \xi \in \mathbb{R}^n, \, t > 0.$$

If we take the inverse Fourier transform with respect to ξ, then we get

$$u(x, t) = (P_t * f)(x), \quad x \in \mathbb{R}^n, \, t > 0,$$

where

$$P_t(x) = (2\pi)^{-n} \int_{\mathbb{R}^n} e^{ix \cdot \xi} e^{-t|\xi|} d\xi, \quad x \in \mathbb{R}^n, \, t > 0.$$

In order to obtain an explicit formula for $P_t(x)$ for all x in $\mathbb{R}^n$ and $t > 0$, we use the following lemma.

DOI: 10.1201/9781003206781-11

Lemma 11.1 *Let β be a positive number. Then*

$$e^{-\beta} = \frac{1}{\sqrt{\pi}} \int_0^\infty \frac{e^{-s}}{\sqrt{s}} e^{-\beta^2/(4s)} ds.$$

Assuming Lemma 11.1 for a moment, and using the third formula in Proposition 4.5 and Proposition 4.6, we get

$$
\begin{aligned}
P_t(x) &= (2\pi)^{-n} \int_{\mathbb{R}^n} e^{ix\cdot\xi} \left(\frac{1}{\sqrt{\pi}} \int_0^\infty \frac{e^{-s}}{\sqrt{s}} e^{-t^2|\xi|^2/(4s)} ds \right) d\xi \\
&= \frac{1}{\sqrt{\pi}} (2\pi)^{-n} \int_0^\infty \frac{e^{-s}}{\sqrt{s}} \left(\int_{\mathbb{R}^n} e^{ix\cdot\xi} e^{-t^2|\xi|^2/(4s)} d\xi \right) ds \\
&= \frac{1}{\sqrt{\pi}} (2\pi)^{-n/2} \int_0^\infty \frac{e^{-s}}{\sqrt{s}} t^{-n} (2s)^{n/2} e^{-s|x|^2/t^2} ds \\
&= 2^{n/2} t^{-n} \frac{1}{\sqrt{\pi}} (2\pi)^{-n/2} \int_0^\infty e^{-s} s^{(n+1)/2} e^{-s|x|^2/t^2} \frac{ds}{s} \\
&= t^{-n} \frac{1}{\pi^{(n+1)/2}} \int_0^\infty e^{-s} s^{(n+1)/2} \frac{ds}{s} \left(1 + \frac{|x|^2}{t^2} \right)^{-(n+1)/2} \\
&= \frac{1}{\pi^{(n+1)/2}} \Gamma\left(\frac{n+1}{2} \right) \frac{t}{(t^2 + |x|^2)^{(n+1)/2}} \\
&= \frac{2}{|\mathbb{S}^n|} \frac{t}{(t^2 + |x|^2)^{(n+1)/2}}
\end{aligned}
$$

for all x in $\mathbb{R}^n$ and $t > 0$. Therefore

$$u(x,t) = (P_t * f)(x), \quad x \in \mathbb{R}^n, \, t > 0,$$

where

$$P_t(x) = \frac{2}{|\mathbb{S}^n|} \frac{t}{(t^2 + |x|^2)^{(n+1)/2}}$$

for all x in $\mathbb{R}^n$ and $t > 0$. The function $P_t(x)$, $x \in \mathbb{R}^n$, $t > 0$, is the Poisson kernel for the Laplacian on the upper half space.

It remains to prove Lemma 11.1. We need another lemma.

Lemma 11.2 *Let β be a positive number. Then*

$$\int_{-\infty}^\infty \frac{e^{ix\beta}}{1 + x^2} dx = \pi e^{-\beta}, \quad \beta > 0.$$

Proof We integrate the function $\frac{e^{iz\beta}}{1+z^2}$ along the boundary C_R of the semidisk with center at the origin and radius R, where $R > 1$. Then using the residue theorem in complex analysis, we get

$$2\pi i \operatorname{Res}\left\{ \frac{e^{iz\beta}}{1 + z^2}; i \right\} = \int_{C_R} \frac{e^{iz\beta}}{1 + z^2} dz = \int_{-R}^R \frac{e^{ix\beta}}{1 + x^2} dx + \int_{\gamma_R} \frac{e^{iz\beta}}{1 + z^2} dz,$$

where γ_R is the circular boundary of the semidisk. Since

$$\left| \int_{\gamma_R} \frac{e^{iz\beta}}{1+z^2} dz \right| \leq \int_0^\pi \frac{e^{-\beta R \sin \theta}}{R^2 - 1} R \, d\theta \to 0$$

as $R \to \infty$, it follows that

$$\int_{-\infty}^\infty \frac{e^{ix\beta}}{1+x^2} dx = 2\pi i \operatorname{Res} \left\{ \frac{e^{iz\beta}}{1+z^2}; i \right\} = 2\pi i \lim_{z \to i} (z - i) \frac{e^{iz\beta}}{1+z^2} = \pi e^{-\beta},$$

as required. □

Proof of Lemma 11.1 Using Lemma 11.2 and the fact that

$$\frac{1}{1+x^2} = \int_0^\infty e^{-(1+x^2)s} ds, \quad x \in \mathbb{R},$$

we get

$$\begin{aligned}
e^{-\beta} &= \frac{1}{\pi} \int_{-\infty}^\infty e^{ix\beta} \left(\int_0^\infty e^{-(1+x^2)s} ds \right) dx \\
&= \frac{1}{\pi} \int_0^\infty e^{-s} \left(\int_{-\infty}^\infty e^{ix\beta} e^{-x^2 s} dx \right) ds \\
&= \frac{1}{\sqrt{\pi}} \int_0^\infty \frac{e^{-s}}{\sqrt{s}} e^{-\beta^2/4s} ds
\end{aligned}$$

if we use the fact that

$$\frac{1}{\sqrt{2\pi}} \int_{-\infty}^\infty e^{ix\beta} e^{-x^2 s} dx = \frac{1}{\sqrt{2s}} e^{-\beta^2/(4s)}.$$

□

That the function u on $\mathbb{R}^n \times (0, \infty)$ defined by

$$u(x,t) = (P_t * f)(x), \quad x \in \mathbb{R}^n, t > 0,$$

is really a solution of the Laplace equation in (11.1) follows from the simple observation that

$$P_t(x) = (2\pi)^{-n} \int_{\mathbb{R}^n} e^{ix \cdot \xi} e^{-t|\xi|} d\xi, \quad x \in \mathbb{R}^n, t > 0,$$

and hence

$$\left(\frac{\partial^2}{\partial t^2} + \Delta \right) (P_t(x)) = 0, \quad x \in \mathbb{R}^n, t > 0.$$

To see that

$$u(x,0) = f(x), \quad x \in \mathbb{R}^n,$$

we note that

$$P_0(x) = (2\pi)^{-n} \int_{\mathbb{R}^n} e^{ix \cdot \xi} d\xi, \quad x \in \mathbb{R}^n.$$

So, by Exercise 8 in Chapter 5, $P_0 = \delta$. Therefore, by Theorem 5.10, we get

$$u(x, 0) = (P_0 * f)(x) = f(x), \quad x \in \mathbb{R}^n.$$

What happens if the function f in the boundary condition in (11.1) is not a Schwartz function? We give the following theorem when $f \in L^p(\mathbb{R}^n)$, $1 \le p < \infty$. The case when $f \in L^\infty(\mathbb{R}^n)$ is an exercise in this chapter.

Theorem 11.3 *Let $f \in L^p(\mathbb{R}^n)$, $1 \le p < \infty$. Then the function u on $\mathbb{R}^n \times (0, \infty)$ defined by*

$$u(x, t) = (P_t * f)(x), \quad x \in \mathbb{R}^n, t > 0,$$

satisfies the boundary condition in (11.1) in the sense that

$$P_t * f \to f$$

in $L^p(\mathbb{R}^n)$ as $t \to 0+$.

Proof Let φ be the function on $\mathbb{R}^n$ defined by

$$\varphi(x) = \frac{\Gamma((n+1)/2)}{\pi^{(n+1)/2}} \frac{1}{(1 + |x|^2)^{(n+1)/2}}, \quad x \in \mathbb{R}^n.$$

Then $\{P_t : t > 0\}$ is the Friedrich mollifier $\{\varphi_t : t > 0\}$ associated to φ, i.e.,

$$P_t(x) = \varphi_t(x) = t^{-n} \varphi\left(\frac{x}{t}\right), \quad x \in \mathbb{R}^n, t > 0.$$

Note that for $t > 0$, we can use the Fourier inversion formula in $\mathcal{S}$ to get

$$
\begin{aligned}
\int_{\mathbb{R}^n} P_t(x) \, dx &= (2\pi)^{-n/2} \int_{\mathbb{R}^n} e^{-ix \cdot 0} \left((2\pi)^{-n/2} \int_{\mathbb{R}^n} e^{ix \cdot \xi} e^{-t|\xi|} d\xi \right) dx \\
&= e^{-t|\xi|} \Big|_{\xi=0} = 1.
\end{aligned}
$$

So, by Theorem 3.4,

$$P_t * f = \varphi_t * f \to f$$

in $L^p(\mathbb{R}^n)$ as $t \to 0+$. $\qquad\square$

Remark 11.4 The construction of the Poisson kernel in this chapter is based on the computation of the inverse Fourier transform of $e^{-t|\xi|}$, $\xi \in \mathbb{R}^n$, $t > 0$, given in the book [41].

Historical Notes

(Peter Gustav) Lejeune Dirichlet (1805–1859) was a German mathematician whose primary interest was in number theory. He was sent to study at the University of Cologne in 1819. Then he was sent to study at Collegè de France in Paris. He presented his first paper to the French Academy of Sciences in 1825. He was appointed Professor at the University of Berlin when he was 23 years old. Besides his interests in number theory, he was interested in analysis and partial differential equations. The Dirichlet kernel in Fourier series and the Dirichlet problem in partial differential equations are named in deference to him. The L-series and L-functions in number theory are also named after him.

Exercises

1. Is the Poisson kernel P_t a tempered function on $\mathbb{R}^n$ for all $t > 0$?

2. Let $f \in L^p(\mathbb{R}^n)$, $1 \leq p \leq \infty$. Prove that the function u defined on the upper half space $\{(x, t) : x \in \mathbb{R}^n, \, t > 0\}$ by

$$u(x, t) = (P_t * f)(x)$$

   is a solution of the partial differential equation in (11.1).

3. Let f be a bounded and continuous function on $\mathbb{R}^n$. Prove that the function u on $\mathbb{R}^n \times (0, \infty)$ defined by

$$u(x, t) = (P_t * f)(x), \quad x \in \mathbb{R}^n, \, t > 0,$$

   satisfies the boundary condition in (11.1) in the sense that

$$P_t * f \to f$$

   uniformly on all compact subsets of $\mathbb{R}^n$ as $t \to 0+$.

4. Find a solution of the Dirichlet problem

$$\begin{cases} \frac{\partial^2 u}{\partial t^2}(x, t) + (\Delta u)(x, t) = 0, & x \in \mathbb{R}^n, \, t > 0, \\ u(\cdot, 0) = \delta. \end{cases}$$

5. Is the solution that you have found for the preceding exercise unique?

Chapter 12

The Bessel–Poisson Kernel

We are now interested in finding the analog of the Poisson kernel for the Dirichlet problem for the perturbed Laplacian on the upper half space $\{(x,t) : x \in \mathbb{R}^n, t > 0\}$ given by

$$\begin{cases} \frac{\partial^2 u}{\partial t^2}(x,t) + ((\Delta - m^2)u)(x,t) = 0, & x \in \mathbb{R}^n, t > 0, \\ u(x,0) = f(x), & x \in \mathbb{R}^n, \end{cases} \tag{12.1}$$

where m is a positive constant and f is a function in $\mathcal{S}$. It turns out that the resulting kernel $W_t(x)$, $x \in \mathbb{R}^n$, $t > 0$, is the product of the Poisson kernel and a weighted Bessel potential.

We again begin with taking the partial Fourier transform with respect to x. Thus, we get

$$\begin{cases} \frac{\partial^2 \hat{u}}{\partial t^2}(\xi,t) - (m^2 + |\xi|^2)\hat{u}(\xi,t) = 0, & \xi \in \mathbb{R}^n, t > 0, \\ \hat{u}(\xi,0) = \hat{f}(\xi), & \xi \in \mathbb{R}^n. \end{cases}$$

As in the case of the Dirichlet problem for the Laplacian on the upper half space, we get

$$\hat{u}(\xi,t) = \hat{f}(\xi)e^{-t\sqrt{m^2+|\xi|^2}}, \quad \xi \in \mathbb{R}^n, t > 0.$$

Taking the inverse Fourier transform with respect to ξ, we get

$$u(x,t) = (W_t * f)(x), \quad x \in \mathbb{R}^n, t > 0,$$

where

$$W_t(x) = (2\pi)^{-n} \int_{\mathbb{R}^n} e^{ix\cdot\xi} e^{-t\sqrt{m^2+|\xi|^2}} d\xi, \quad x \in \mathbb{R}^n, t > 0.$$

DOI: 10.1201/9781003206781-12

In order to obtain an explicit formula for $W_t(x)$, $x \in \mathbb{R}^n$, $t > 0$, we again employ Lemma 11.1 and we get

$$
\begin{aligned}
W_t(x) &= \frac{(2\pi)^{-n}}{\sqrt{\pi}} \int_{\mathbb{R}^n} e^{ix\cdot\xi} \left(\int_0^\infty \frac{e^{-s}}{\sqrt{s}} e^{-(m^2+|\xi|^2)t^2/(4s)} ds \right) d\xi \\
&= \frac{(2\pi)^{-n}}{\sqrt{\pi}} \int_0^\infty \frac{e^{-s}}{\sqrt{s}} e^{-m^2t^2/(4s)} \left(\int_{\mathbb{R}^n} e^{ix\cdot\xi} e^{-|\xi|^2t^2/(4s)} d\xi \right) ds \\
&= (2\pi)^{-n/2} \frac{1}{\sqrt{\pi}} \int_0^\infty \frac{e^{-s}}{\sqrt{s}} e^{-m^2t^2/(4s)} (2s)^{n/2} t^{-n} e^{-s|x|^2/t^2} ds \\
&= t^{-n} \pi^{(n+1)/2} \int_0^\infty e^{-s\left(1+\frac{|x|^2}{t^2}\right)} e^{-m^2t^2/(4s)} s^{(n+1)/2} \frac{ds}{s} \\
&= \frac{1}{\pi^{(n+1)/2}} \frac{t}{(t^2+|x|^2)^{(n+1)/2}} \int_0^\infty e^{-r} e^{-m^2(t^2+|x|^2)/(4r)} r^{(n+1)/2} \frac{dr}{r}
\end{aligned}
$$

for all x in $\mathbb{R}^n$ and $t > 0$. Therefore

$$
u(x,t) = (W_t * f)(x), \quad x \in \mathbb{R}^n, t > 0,
$$

where

$$
W_t(x) = \frac{t}{(t^2+|x|^2)^{(n+1)/2}} \frac{1}{\pi^{(n+1)/2}} \int_0^\infty e^{-r} e^{-m^2(t^2+|x|^2)/(4r)} r^{(n+1)/2} \frac{dr}{r}
$$

for all $x \in \mathbb{R}^n$ and $t > 0$. It is worthwhile to observe that if $m = 0$, then

$$
W_t(x) = P_t(x), \quad x \in \mathbb{R}^n, t > 0,
$$

which is the result in the preceding chapter.

We can go one step further to obtain a more succinct formula for the kernel $W_t(x)$, $x \in \mathbb{R}^n$, $t > 0$. For the sake of notation, we define for all positive numbers s, m, and t, the weighted Bessel potential $G_{s,m,t}$ on $\mathbb{R}^n$ by

$$
G_{s,m,t}(x) = \frac{m^{n-s}}{2^{s/2}\Gamma(s/2)} \int_0^\infty e^{-r/2} e^{-m^2(t^2+|x|^2)/(2r)} r^{-(n-s)/2} \frac{dr}{r}, \quad x \in \mathbb{R}^n.
$$

Remark 12.1 In light of the Lebesgue dominated convergence theorem, it is easy to see that for all x in $\mathbb{R}^n \setminus \{0\}$,

$$
G_{s,m,t}(x) \to G_{s,m}(x)
$$

as $t \to 0+$. Furthermore, if $s > m$, then

$$
G_{s,m,t} \to G_{s,m}
$$

uniformly on $\mathbb{R}^n$ as $t \to 0+$. Indeed,

$$
\begin{aligned}
&\sup_{x \in \mathbb{R}^n} |G_{s,m,t}(x) - G_{s,m}(x)| \\
&= \frac{m^{n-s}}{2^{s/2}\Gamma(s/2)} \sup_{x \in \mathbb{R}^n} \left| \int_{\mathbb{R}^n} e^{-r/2}(e^{m^2t^2/(2r)} - 1)e^{-m^2|x|^2/(2r)} r^{-(n-s)/2} \frac{dr}{r} \right| \\
&\to 0.
\end{aligned}
$$

Now, we let $r = \frac{\delta}{2}$ in the formula for $W_t(x)$, $x \in \mathbb{R}^n$, $t > 0$. Then

$$W_t(x)$$

$$= \frac{t}{(t^2 + |x|^2)^{(n+1)/2}} \frac{1}{(2\pi)^{(n+1)/2}} \int_0^\infty e^{-\delta/2} e^{-m^2(t^2+|x|^2)/(2\delta)} \delta^{(n+1)/2} \frac{d\delta}{\delta}$$

$$= \gamma_{m,n} P_t(x) G_{2n+1,m,t}(x)$$

for all x in $\mathbb{R}^n$ and $t > 0$, where

$$\gamma_{m,n} = 2^{n/2} m^{n+1} \frac{\Gamma((2n+1)/2)}{\Gamma((n+1)/2)}.$$

Thus, the analog of the Poisson kernel for the Dirichlet problem for the perturbed Laplacian on the upper half space is the product of the Poisson kernel and the weighted Bessel potential $G_{2n+1,m,t}$. We call it the Bessel–Poisson kernel.

We leave it as an exercise to prove that for every function f in $L^p(\mathbb{R}^n)$, $1 \leq p \leq \infty$, the function u on $\mathbb{R}^n \times (0, \infty)$ defined by

$$u(x,t) = (W_t * f)(x), \quad x \in \mathbb{R}^n \times (0, \infty),$$

is a solution of the partial differential equation in (12.1).

As for the boundary condition in (12.1), we have the following result.

Theorem 12.2 *Let $f \in L^p(\mathbb{R}^n)$, $1 \leq p < \infty$. Then*

$$W_t * f \to f$$

in $L^p(\mathbb{R}^n)$ as $t \to 0+$.

Proof We have

$$(W_t * f)(x) - f(x)$$

$$= \int_{\mathbb{R}^n} \gamma_{m,n} P_t(y) G_{2n+1,m,t}(y) f(x-y) \, dy - \int_{\mathbb{R}^n} P_t(y) f(x) \, dy$$

$$= I_t(x) + J_t(x)$$

for all $x \in \mathbb{R}^n$ and $t > 0$, where

$$I_t(x) = \int_{\mathbb{R}^n} P_t(y) \gamma_{m,n} (G_{2n+1,m,t}(y) - G_{2n+1,m}(y)) f(x-y) \, dy$$

and

$$J_t(x) = \int_{\mathbb{R}^n} P_t(y) \gamma_{m,n} G_{2n+1,m}(y) f(x-y) \, dy - \int_{\mathbb{R}^n} P_t(y) f(x) \, dy.$$

Now, by Minkowski's inequality in integral form, the fact that

$$\int_{\mathbb{R}^n} P_t(y) \, dy = 1, \quad t > 0,$$

and Remark 12.1,

$$\|I_t\|_p$$

$$= \gamma_{m,n} \left(\int_{\mathbb{R}^n} \left| \int_{\mathbb{R}^n} P_t(y)(G_{2n+1,m,y}(y) - G_{2n+1,m}(y))f(x-y)\,dy \right|^p dx \right)^{1/p}$$

$$\leq \gamma_{m,n} \int_{\mathbb{R}^n} \left(\int_{\mathbb{R}^n} |P_t(y)(G_{2n+1,m,t}(y) - G_{2n+1,m}(y))f(x-y)|^p dx \right)^{1/p} dy$$

$$= \gamma_{m,n} \int_{\mathbb{R}^n} P_t(y)|G_{2n+1,m,t}(y) - G_{2n+1,m}(y)|\,dy\,\|f\|_p$$

$$\leq \gamma_{m,n} \sup_{y \in \mathbb{R}^n} |G_{2n+1,m,t}(y) - G_{2n+1,m}(y)| \int_{\mathbb{R}^n} P_t(y)\,dy\,\|f\|_p$$

$$\leq \gamma_{m,n} \sup_{y \in \mathbb{R}^n} |G_{2n+1,m,t}(y) - G_{2n+1,m}(y)|\,\|f\|_p \to 0$$

as $t \to 0+$. For all $x \in \mathbb{R}^n$ and $t > 0$, we have

$$J_t(x) = L_t(x) + M_t(x),$$

where

$$L_t(x) = \int_{\mathbb{R}^n} P_t(y)(\gamma_{m,n} G_{2n+1,m}(y) - 1)f(x-y)\,dy$$

and

$$M_t(x) = \int_{\mathbb{R}^n} P_t(y)(f(x-y) - f(x))\,dy.$$

Note that, by Theorem 11.3,

$$\|M_t\|_p = \|P_t * f - f\|_p \to 0$$

as $t \to 0+$. Now, as in the analysis of I_t, we can use Minkowski's inequality in integral form to get

$$\|L_t\|_p \leq \int_{\mathbb{R}^n} P_t(y)|\gamma_{m,n} G_{2n+1,m}(y) - 1|\,dy\,\|f\|_p, \quad t > 0.$$

Since

$$\gamma_{m,n} G_{2n+1,m}(0) = 1,$$

it follows from Theorem 3.12 that

$$\int_{\mathbb{R}^n} P_t(y)|\gamma_{m,n} G_{2n+1,m}(y) - 1|\,dy \to |\gamma_{m,n} G_{2n+1,m}(0) - 1| = 0$$

as $t \to 0+$. Thus, $\|L_t\|_p \to 0$ as $t \to 0+$. So,

$$\|J_t\|_p \leq \|L_t\|_p + \|M_t\|_p \to 0$$

as $t \to 0+$. Therefore

$$\|W_t * f - f\|_p \leq \|I_t\|_p + \|J_t\|_p \to 0$$

as $t \to 0+$ and the proof is complete. $\qquad\qquad\qquad\qquad\qquad\qquad\qquad\qquad$ □

Exercises

1. Let $f \in L^p(\mathbb{R}^n)$, $1 \leq p \leq \infty$. Prove that the function u on $\mathbb{R}^n \times (0, \infty)$ defined by
$$u(x, t) = (W_t * f)(x), \quad x \in \mathbb{R}^n, \, t > 0,$$
is a solution of the Dirichlet problem (12.1).

2. Prove that if $f \in L^\infty(\mathbb{R}^n)$ is a continuous function on $\mathbb{R}^n$, then
$$W_t * f \to f$$
uniformly on every compact subset of $\mathbb{R}^n$ as $t \to 0+$.

3. Prove that $W_t \in L^p(\mathbb{R}^n)$, $1 \leq p \leq \infty$, for all $t > 0$.

4. Find a bounded solution of the Dirichlet problem
$$\begin{cases} \frac{\partial^2 u}{\partial t^2}(x, t) + ((\Delta - m^2)u)(x, t) = 0, & x \in \mathbb{R}^n, \, t > 0, \\ u(\cdot, 0) = \delta. \end{cases}$$

5. Is the solution that you have found for the preceding exercise the unique bounded solution?

Chapter 13

Wave Kernels

Let us look at the initial value problem for the wave equation given by

$$\begin{cases} \frac{\partial^2 u}{\partial t^2}(x,t) = (\Delta u)(x,t), & x \in \mathbb{R}^n, \, t > 0, \\ u(x,0) = f(x), & x \in \mathbb{R}^n, \\ \frac{\partial u}{\partial t}(x,0) = g(x), & x \in \mathbb{R}^n, \end{cases} \tag{13.1}$$

where f and g are given functions in $\mathcal{S}$.

If we write the Laplace equation in Chapter 11 as

$$\frac{\partial^2 u}{\partial t^2}(x,t) = -(\Delta u)(x,t), \quad x \in \mathbb{R}^n, \, t > 0,$$

then we see that the wave equation differs from the Laplace equation by the opposite sign in front of the Laplacian Δ. Again, the two equations are dramatically different in the sense that the solution of the Laplace equation in Chapter 11 describes potential theory and the solution of the wave equation gives wave propagation.

We begin with taking the partial Fourier transform of u with respect to x and get

$$\begin{cases} \frac{\partial^2 \hat{u}}{\partial t^2}(\xi,t) + |\xi|^2 \hat{u}(\xi,t) = 0, & \xi \in \mathbb{R}^n, \, t > 0, \\ \hat{u}(\xi,0) = \hat{f}(\xi), & \xi \in \mathbb{R}^n, \\ \frac{\partial \hat{u}}{\partial t}(\xi,0) = \hat{g}(\xi), & \xi \in \mathbb{R}^n. \end{cases}$$

Thus, we get

$$\hat{u}(\xi,t) = C_1 \cos(|\xi|t) + C_2 \sin(|\xi|t), \quad \xi \in \mathbb{R}^n, \, t > 0.$$

Using the initial condition for $\hat{u}(\xi,0)$, we get $C_1 = \hat{f}(\xi)$. Since

$$\frac{\partial \hat{u}}{\partial t}(\xi,t) = -C_1|\xi|\sin(|\xi|t) + C_2|\xi|\cos(|\xi|t),$$

it follows from the initial condition for $\frac{\partial \hat{u}}{\partial t}(\xi,0)$ that

$$C_2 = \frac{\hat{g}(\xi)}{|\xi|}, \quad \xi \neq 0.$$

DOI: 10.1201/9781003206781-13

Thus,

$$\hat{u}(\xi,t) = \cos(|\xi|t)\hat{f}(\xi) + \frac{\sin(|\xi|t)}{|\xi|}\hat{g}(\xi), \quad \xi \in \mathbb{R}^n \setminus \{0\}, \, t > 0,$$

and hence if we take the inverse Fourier transform with respect to ξ, then we get

$$u(x,t) = (2\pi)^{-n/2}\int_{\mathbb{R}^n} e^{ix\cdot\xi}\cos(|\xi|t)\hat{f}(\xi)\,d\xi + (2\pi)^{-n/2}\int_{\mathbb{R}^n} e^{ix\cdot\xi}\frac{\sin(|\xi|t)}{|\xi|}\hat{g}(\xi)\,d\xi$$

for all x in $\mathbb{R}^n$ and $t > 0$. This is the Fourier integral representation of the solution of the initial value problem for the wave equation in $\mathbb{R}^n$. Writing out the Fourier transforms explicitly, we get for all x in $\mathbb{R}^n$ and $t > 0$,

$$u(x,t) = (2\pi)^{-n/2}\int_{\mathbb{R}^n} C_t(x-y)\,f(y)\,dy + (2\pi)^{-n/2}\int_{\mathbb{R}^n} S_t(x-y)\,g(y)\,dy,$$

where

$$C_t(x) = (2\pi)^{-n/2}\int_{\mathbb{R}^n} e^{ix\cdot\xi}\cos(|\xi|t)\,d\xi$$

and

$$S_t(x) = (2\pi)^{-n/2}\int_{\mathbb{R}^n} e^{ix\cdot\xi}\frac{\sin(|\xi|t)}{|\xi|}\,d\xi.$$

We call $C_t(x)$ and $S_t(x)$, $x \in \mathbb{R}^n$, $t > 0$, the wave kernels.

As an illustration of the use of the formula for the solution that we have just obtained, we derive the d'Alembert formula for the solution in one dimension. Indeed, using the formula

$$\hat{\chi}(\xi) = \sqrt{\frac{2}{\pi}}\frac{\sin\xi}{\xi}, \quad \xi \in \mathbb{R},$$

where

$$\chi(x) = \begin{cases} 1, & x \in [-1,1], \\ 0, & x \notin [-1,1], \end{cases}$$

and $\frac{\sin\xi}{\xi}$ is understood to be 1 if $\xi = 0$, we get

$$\widehat{D_{\frac{1}{t}}\chi}(\xi) = t\hat{\chi}(t\xi) = \sqrt{\frac{2}{\pi}}\frac{\sin(t\xi)}{\xi}, \quad \xi \in \mathbb{R}, \, t > 0.$$

Therefore the wave kernel S_t in one dimension is given by

$$S_t(x) = \sqrt{\frac{\pi}{2}}(D_{\frac{1}{t}}\chi)(x), \quad x \in \mathbb{R}, \, t > 0.$$

Since

$$((D_{\frac{1}{t}}\chi) * g)^{\wedge}(\xi) = (2\pi)^{1/2}\widehat{D_{\frac{1}{t}}\chi}(\xi)\hat{g}(\xi) = 2\frac{\sin(t|\xi|)}{|\xi|}\hat{g}(\xi)$$

for all ξ in $\mathbb{R}$ and $t > 0$, we get

$$(2\pi)^{-1/2} \int_{-\infty}^{\infty} e^{ix\xi} \frac{\sin(t|\xi|)}{|\xi|} \hat{g}(\xi) \, d\xi = \frac{1}{2}((D_{\frac{1}{t}}\chi) * g)(x)$$

$$= \frac{1}{2} \int_{-\infty}^{\infty} \chi\left(\frac{x-y}{t}\right) g(y) \, dy = \frac{1}{2} t \int_{-\infty}^{\infty} \chi(z) g(x - tz) \, dz$$

$$= \frac{1}{2} t \int_{-1}^{1} g(x - tz) \, dz = \frac{1}{2} \int_{x-t}^{x+t} g(z) \, dz$$

for all x in $\mathbb{R}$ and $t > 0$. Let us now compute $(2\pi)^{-1/2} \int_{-\infty}^{\infty} e^{ix\xi} \cos(|\xi|t) d\xi$. To this end, we note that

$$\cos(|\xi|t) = \cos(\xi t) = \frac{1}{2}(e^{i\xi t} + e^{-i\xi t}), \quad \xi \in \mathbb{R}, \, t > 0.$$

But, using the fact that the Fourier transform of the one-dimensional Dirac delta δ is $(2\pi)^{-1/2}$ and Exercise 13 in Chapter 5,

$$(2\pi)^{-1/2} e^{i\xi t} = \widehat{T_t \delta}(\xi)$$

and hence

$$e^{i\xi t} = (2\pi)^{1/2} \widehat{T_t \delta}(\xi), \quad \xi \in \mathbb{R}, \, t > 0.$$

Similarly,

$$e^{-i\xi t} = (2\pi)^{1/2} \widehat{T_{-t} \delta}(\xi), \quad \xi \in \mathbb{R}, \, t > 0.$$

We have

$$\cos(|\xi|t) = \sqrt{\frac{\pi}{2}} (\widehat{T_t \delta}(\xi) + \widehat{T_{-t} \delta}(\xi)), \quad \xi \in \mathbb{R}, \, t > 0.$$

So, we get

$$(2\pi)^{-1/2} \int_{-\infty}^{\infty} e^{ix\xi} \cos(|\xi|t) \, d\xi = C_t(x), \quad x \in \mathbb{R}, \, t > 0,$$

where C_t is the other wave kernel in one dimension given by

$$C_t(x) = \sqrt{\frac{\pi}{2}} (T_t \delta + T_{-t} \delta)(x), \quad x \in \mathbb{R}, \, t > 0.$$

Since

$$(C_t * f)^{\wedge}(\xi) = (2\pi)^{1/2} \widehat{C_t}(\xi) \hat{f}(\xi) = (2\pi)^{1/2} \cos(|\xi|t) \hat{f}(\xi)$$

for all ξ in $\mathbb{R}$ and $t > 0$, it follows that

$$(2\pi)^{-1/2} \int_{-\infty}^{\infty} e^{ix\xi} \cos(|\xi|t) \hat{f}(\xi) \, d\xi$$

$$= (2\pi)^{-1/2} \int_{-\infty}^{\infty} C_t(x-y) f(y) \, dy = \frac{1}{2} \int_{-\infty}^{\infty} (T_t \delta + T_{-t}\delta)(x-y) f(y) \, dy$$

$$= \frac{1}{2} \int_{-\infty}^{\infty} (T_t \delta)(x-y) f(y) \, dy + \frac{1}{2} \int_{-\infty}^{\infty} (T_{-t}\delta)(x-y) f(y) \, dy$$

$$= \frac{1}{2} \int_{-\infty}^{\infty} \delta(x-y+t) f(y) \, dy + \frac{1}{2} \int_{-\infty}^{\infty} \delta(x-y-t) f(y) \, dy$$

$$= \frac{1}{2}(f(x+t) + f(x-t)), \quad x \in \mathbb{R}, \, t > 0.$$

Therefore the solution of the initial value problem for the wave equation in one dimension is given by

$$u(x,t) = \frac{1}{2}(f(x+t) + f(x-t)) + \frac{1}{2} \int_{x-t}^{x+t} g(z) \, dz \qquad (13.2)$$

for all x in $\mathbb{R}$ and $t > 0$. If we let G be an antiderivative of g, then

$$u(x,t) = \frac{1}{2}(f(x+t) + f(x-t)) + \frac{1}{2}(G(x+t) - G(x-t)), \quad x \in \mathbb{R}, \, t > 0.$$

Thus, we have the d'Alembert formula for the solution, which is in fact a superposition of traveling waves to the left and to the right.

We can also give the Poisson formula for the solution in $\mathbb{R}^3$. We note that

$$(2\pi)^{-3/2} \int_{\mathbb{R}^3} e^{ix\cdot\xi} \frac{\sin(|\xi|t)}{|\xi|} \hat{g}(\xi) \, d\xi$$

$$= (2\pi)^{-3} \int_{\mathbb{R}^3} \int_{\mathbb{R}^3} e^{i(x-y)\cdot\xi} \frac{\sin(|\xi|t)}{|\xi|} g(y) \, dy \, d\xi$$

$$= (2\pi)^{-3} \int_{\mathbb{R}^3} \left(\int_{\mathbb{R}^3} e^{i(x-y)\cdot\xi} \frac{\sin(|\xi|t)}{|\xi|} \, d\xi \right) g(y) \, dy, \quad x \in \mathbb{R}^3, \, t > 0.$$

Now, we use spherical coordinates (ρ, ψ, τ) in the ξ-space where the north pole $\psi = 0$ is in the direction of $x - y$. We also use spherical coordinates (r, ω) in the y-space with center at x, where $y = x + r\omega$, $|\omega| = 1$. Then

$$(2\pi)^{-3/2} \int_{\mathbb{R}^3} e^{i(x-y)\cdot\xi} \frac{\sin(|\xi|t)}{|\xi|} \, d\xi$$

$$= (2\pi)^{-3/2} \int_0^{2\pi} \int_0^{\pi} \int_0^{\infty} e^{ir\rho\cos\psi} \frac{\sin(\rho t)}{\rho} \rho^2 \sin\psi \, d\rho \, d\psi \, d\tau$$

$$= (2\pi)^{-1/2} \int_0^{\infty} \frac{e^{ir\rho\cos\psi}}{-ir} \Big|_0^{\pi} \sin(\rho t) \, d\rho = (2\pi)^{-1/2} \frac{2}{r} \int_0^{\infty} \sin(r\rho) \sin(\rho t) \, d\rho$$

for all ξ and y in $\mathbb{R}^3$ and $t > 0$. So, for all x in $\mathbb{R}^3$ and $t > 0$,

$$(2\pi)^{-3/2} \int_{\mathbb{R}^3} e^{ix\cdot\xi} \frac{\sin(|\xi|t)}{|\xi|} \hat{g}(\xi)\, d\xi$$
$$= (2\pi)^{-2} 2 \int_0^\infty \sin(\rho t) \left(\int_{\mathbb{S}^2} \int_0^\infty g(x + r\omega)\sin(r\rho)\, r\, dr\, d\sigma(\omega) \right) d\rho.$$

Let $G(r)$ be the function on $\mathbb{R}$ defined to be the odd extension of

$$\int_{\mathbb{S}^2} g(x + r\omega) r\, d\sigma(\omega)$$

to all of $\mathbb{R}$. Then

$$\int_0^\infty \sin(r\rho)\, G(r)\, dr = -\frac{1}{2i} \int_{-\infty}^\infty e^{-ir\rho} G(r)\, dr = -\frac{1}{2i} (2\pi)^{1/2} \widehat{G}(\rho), \quad \rho > 0.$$

Using the Fourier inversion formula, we get

$$(2\pi)^{-3/2} \int_{\mathbb{R}^3} e^{ix\cdot\xi} \frac{\sin(|\xi|t)}{|\xi|} \hat{g}(\xi)\, d\xi$$
$$= (2\pi)^{-2} 2 \left(-\frac{1}{2i} (2\pi)^{1/2} \right) \int_0^\infty \sin(\rho t)\, \widehat{G}(\rho)\, d\rho$$
$$= -\frac{1}{i} (2\pi)^{-3/2} \frac{1}{2i} \int_{-\infty}^\infty e^{i\rho t} \widehat{G}(\rho)\, d\rho$$
$$= \frac{1}{2} (2\pi)^{-3/2} \int_{-\infty}^\infty e^{i\rho t} \widehat{G}(\rho)\, d\rho$$
$$= \frac{1}{2} (2\pi)^{-1} G(t)$$
$$= \frac{1}{4\pi} \int_{\mathbb{S}^2} g(x + t\omega) t\, d\sigma(\omega)$$
$$= \frac{t}{|\mathbb{S}^2|} \int_{\mathbb{S}^2} g(x + t\omega)\, d\sigma(\omega)$$

for all x in $\mathbb{R}^3$ and $t > 0$. This is t times the average of the initial data g on the sphere of radius t centered at x. We now let p and P be the functions on $\mathbb{R}^3 \times (0, \infty)$ defined, respectively, by

$$p(x,t) = (2\pi)^{-3/2} \int_{\mathbb{R}^3} e^{ix\cdot\xi} \cos(|\xi|t) \hat{f}(\xi)\, d\xi$$

and

$$P(x,t) = \int_0^t p(x,s)\, ds$$

for all x in $\mathbb{R}^3$ and $t > 0$. Then, by what we have obtained for g, we get

$$
\begin{aligned}
P(x,t) &= (2\pi)^{-3/2} \int_{\mathbb{R}^3} e^{ix\cdot\xi}\, \frac{\sin(|\xi|s)}{|\xi|} \bigg|_{s=0}^{s=t} \hat{f}(\xi)\, d\xi \\
&= (2\pi)^{-3/2} \int_{\mathbb{R}^3} e^{ix\cdot\xi}\, \frac{\sin(|\xi|t)}{|\xi|} \hat{f}(\xi)\, d\xi \\
&= \frac{t}{|S^2|} \int_{S^2} f(x + t\omega)\, d\sigma(\omega)
\end{aligned}
$$

for all x in $\mathbb{R}^3$ and $t > 0$. Hence the solution of the initial value problem for the wave equation in $\mathbb{R}^3$ is given by

$$
u(x,t) = \frac{\partial}{\partial t}\left\{ \frac{t}{|S^2|} \int_{S^2} f(x + t\omega)\, d\sigma(\omega)\right\} + \frac{t}{|S^2|} \int_{S^2} g(x + t\omega)\, d\sigma(\omega)
$$

for all x in $\mathbb{R}^3$ and $t > 0$. This is the Poisson formula.

We now turn our attention to the initial value problem for the wave equation in $\mathbb{R}^2$ given by

$$
\begin{cases}
\frac{\partial^2 u}{\partial t^2}(x,t) = (\Delta u)(x,t), & x \in \mathbb{R}^2, t > 0, \\
u(x,0) = f(x), & x \in \mathbb{R}^2, \\
\frac{\partial u}{\partial t}(x,0) = g(x), & x \in \mathbb{R}^2,
\end{cases}
$$

where f and g are functions in $\mathcal{S}$.

To solve the initial value problem in $\mathbb{R}^2$, we think of the functions f and g as functions of (x, x_3) on $\mathbb{R}^3$, which do not depend on x_3, and make use of the Poisson formula for the solution of the initial value problem in $\mathbb{R}^3$. Thus, the solution

$$
v = v(x, x_3, t)
$$

of

$$
\begin{cases}
\frac{\partial^2 v}{\partial t^2}(x, x_3, t) = (\Delta v)(x, x_3, t), & (x, x_3) \in \mathbb{R}^3, t > 0, \\
v(x, x_3, 0) = f(x), & (x, x_3) \in \mathbb{R}^3, \\
\frac{\partial v}{\partial t}(x, x_3, 0) = g(x), & (x, x_3) \in \mathbb{R}^3,
\end{cases}
$$

is given by

$$
\begin{aligned}
v(x, x_3, t) &= \frac{\partial}{\partial t}\left\{ \frac{t}{|S^2|} \int_{|y|^2 + y_3^2 = 1} f(x + t\sigma(y, y_3))\, d\sigma(y, y_3)\right\} \\
&\quad + \frac{t}{|S^2|} \int_{|y|^2 + y_3^2 = 1} g(x + t\sigma(y, y_3))\, d\sigma(y, y_3)
\end{aligned}
$$

for all (x, x_3) in $\mathbb{R}^3$ and $t > 0$. Obviously, $v(x, x_3, t)$ is independent of x_3, and hence we can write

$$
v(x, x_3, t) = u(x, t), \quad (x, x_3) \in \mathbb{R}^3, t > 0.
$$

Therefore

$$\frac{\partial u}{\partial t} = \frac{\partial v}{\partial t} = \Delta v = \sum_{j=1}^{3} \frac{\partial^2 v}{\partial x_j^2} = \sum_{j=1}^{2} \frac{\partial^2 u}{\partial x_j^2} = \Delta u$$

and hence $u(x,t)$ is the solution of the initial value problem in $\mathbb{R}^2$ for all x in $\mathbb{R}^2$ and $t > 0$. This method of solving the initial value problem in $\mathbb{R}^2$ by descending from the corresponding solution of the initial value problem in $\mathbb{R}^3$ is known as the method of descent.

Let us now examine the solution of the initial value problem in $\mathbb{R}^2$ obtained by the method of descent in some detail. We first give a lemma.

Lemma 13.1 *The surface measure on the unit sphere*

$$\{(y, y_3) \in \mathbb{R}^3 : |y|^2 + y_3^2 = 1\}$$

centered at the origin is given in yy_3-coordinates by

$$d\sigma(y, y_3) = \frac{dy}{\sqrt{1 - |y|^2}}, \quad |y| \leq 1.$$

Proof Since $|y|^2 + y_3^2 = 1$, it follows that

$$y_3 = \sqrt{1 - |y|^2}, \quad |y| \leq 1.$$

So the surface measure on the unit sphere centered at the origin is given by

$$\begin{aligned}
d\sigma(y, y_3) &= \sqrt{1 + \left(\frac{\partial y_3}{\partial y_1}\right)^2 + \left(\frac{\partial y_3}{\partial y_2}\right)^2}\, dy \\
&= \sqrt{1 + \frac{y_1^2}{1 - |y|^2} + \frac{y_2^2}{1 - |y|^2}}\, dy = \frac{1}{\sqrt{1 - |y|^2}} dy.
\end{aligned}$$

$\square$

We can now use Lemma 13.1 to write

$$\begin{aligned}
&u(x,t) \\
&= \frac{\partial}{\partial t} \left\{ \frac{t}{|\mathbb{S}^2|} \int_{|y|^2 + y_3^2 = 1} f(x + t\sigma(y, y_3))\, d\sigma(y, y_3) \right\} \\
&\quad + \frac{t}{|\mathbb{S}^2|} \int_{|y|^2 + y_3^2 = 1} g(x + t\sigma(y, y_3))\, d\sigma(y, y_3) \\
&= \frac{\partial}{\partial t} \left\{ \frac{t}{|\mathbb{S}^2|} \int_{|y| \leq 1} f(x + ty) \frac{dy}{\sqrt{1 - |y|^2}} \right\} + \frac{t}{|\mathbb{S}^2|} \int_{|y| \leq 1} g(x + ty) \frac{dy}{\sqrt{1 - |y|^2}}
\end{aligned}$$

for all x in $\mathbb{R}^2$ and $t > 0$.

A comparison of the solution in $\mathbb{R}^2$ with that in $\mathbb{R}^3$ is interesting. In $\mathbb{R}^2$, the value $u(x_0, t_0)$ of the solution u at a point (x_0, t_0) in $\mathbb{R}^2 \times (0, \infty)$ depends on all the values of the initial data on the entire disk with center x_0 and radius t_0. But in $\mathbb{R}^3$, the value $u(x_0, t_0)$ of the solution u at a point (x_0, t_0) in $\mathbb{R}^3 \times (0, \infty)$ depends only on the values of the initial data on the surface of the sphere $\mathbb{S}(x_0, t_0)$ with center x_0 and radius t_0. This fact is known as the Huygens principle.

As an instructive illustration of the Huygens principle, suppose that a very localized sound occurs at the center of the universe, which we assume is $\mathbb{R}^3$. What is the sound effect on an observer located at the point x in $\mathbb{R}^3$ as time t evolves? At the beginning when $t < |x|$, the sphere with center at x and radius t fails to touch the sound and hence the observer cannot hear anything. When $t = |x|$, the observer hears the sound with magnitude given by $|x|$ times the average of the source over the sphere with center at x and radius $|x|$. When $t > |x|$, the observer again hears nothing. The observer hears the sound emanating from the source just at time $t = |x|$. See FIGURE 13.1.

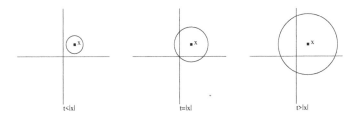

FIGURE 13.1: The spheres of observance.

To contrast the sound effect just described, let us look at the same set up for the universe $\mathbb{R}^2$. Again at the beginning when $t < |x|$, the observer's sphere (= circle) of "observance" falls short of the source and therefore nothing can be heard. When $t = |x|$, the observer begins to hear a sound and will continue to hear the sound when $t > |x|$. The intensity of the sound will go to 0 as $t \to \infty$, but is there at any finite time $t > |x|$. The following FIGURE 13.2 illustrates this phenomenon.

FIGURE 13.2: The sound effect.

Remark 13.2 The constructions of the solutions of the solutions of the initial value problem for the wave equation in $\mathbb{R}^3$ and $\mathbb{R}^2$ in this chapter are similar to the corresponding constructions in the book [51].

Historical Notes

Jean le Rond d'Alembert (1717–1783), a French mathematician and physicist, was a contemporary of Euler. In mathematics, he is best remembered for the d'Alembert principle in mechanics and the d'Alembert formula for the solution of the wave equation that we have seen in this chapter.

Christiaan Huygens (1629–1695), a Dutch physicist and astronomer, is best remembered for his works on the understanding the pendulum, the invention of the pendulum clock, and the initial development of a wave theory of light. The discoveries of the rings of Saturn and its satellite Titan are also attributed to him.

Exercises

1. Prove that the function u on $\mathbb{R} \times (0, \infty)$ defined by (13.2) is indeed a solution of the initial value problem (13.1).

2. Let u be the solution of the initial value problem

$$\begin{cases} \frac{\partial^2 u}{\partial t^2}(x,t) = (\Delta u)(x,t), & x \in \mathbb{R}^3,\, t > 0, \\ u(x,0) = 0, & x \in \mathbb{R}^3, \\ \frac{\partial u}{\partial t}(x,0) = |x|^2, & x \in \mathbb{R}^3, \end{cases}$$

   obtained by the method in this chapter. Find $u(0,t)$ for $t > 0$.

3. Use the method in this chapter to solve the initial value problem

$$\begin{cases} \frac{\partial^2 u}{\partial t^2}(x,t) = (\Delta u)(x,t), & x \in \mathbb{R}^3,\, t > 0, \\ u(x,0) = 0, & x \in \mathbb{R}^3, \\ \frac{\partial u}{\partial t}(x,0) = x_1, & x \in \mathbb{R}^3. \end{cases}$$

4. Do Exercise 1 with $\mathbb{R}^3$ replaced by $\mathbb{R}^2$.

5. Do Exercise 2 with $\mathbb{R}^3$ replaced by $\mathbb{R}^2$.

Chapter 14

The Heat Kernel of the Hermite Operator

The rest of the book is devoted to a study of some partial differential operators with variable coefficients. As in preceding chapters, the emphasis is on important operators arising in physics. We begin in this chapter with the Hermite operator, which is a mathematical model in physics. First we give a recall without proofs of the basic properties of Hermite functions that we need in this chapter and to some extent the following chapters. The books [47, 54] are useful references.

For $k = 0, 1, 2, \ldots$, the Hermite function of order k is the function e_k on $\mathbb{R}$ defined by

$$e_k(x) = \frac{1}{(2^k k! \sqrt{\pi})^{1/2}} e^{-x^2/2} H_k(x), \quad x \in \mathbb{R}, \tag{14.1}$$

where H_k is the Hermite polynomial of degree k given by

$$H_k(x) = (-1)^k e^{x^2} \left(\frac{d}{dx}\right)^k (e^{-x^2}), \quad x \in \mathbb{R}.$$

It is well known that $\{e_k : k = 0, 1, 2, \ldots\}$ is an orthonormal basis for $L^2(\mathbb{R})$.

Remark 14.1 There exists a constant K such that

$$|e_k(x)| \leq K\pi^{-1/4}, \quad x \in \mathbb{R},$$

for $k = 0, 1, 2, \ldots$. Moreover,
$$K < 1.09.$$

This inequality is known as Cramér's inequality. This inequality and its references can be found in the paper [16].

Let A and A^* be differential operators on $\mathbb{R}$ defined by

$$A = \frac{d}{dx} + x$$

DOI: 10.1201/9781003206781-14

and

$$A^* = -\frac{d}{dx} + x.$$

In fact, A^* is the formal adjoint of A in the sense that

$$\int_{-\infty}^{\infty} (Au)(x)\overline{v(x)}\,dx = \int_{-\infty}^{\infty} u(x)\overline{(A^*v)(x)}\,dx$$

for all Schwartz functions u and v on $\mathbb{R}$. The one-dimensional Hermite operator H is the ordinary differential operator on $\mathbb{R}$ given by

$$H = \frac{1}{2}(AA^* + A^*A).$$

A simple calculation shows that

$$H = -\frac{d^2}{dx^2} + x^2.$$

The spectral analysis of H is based on the following result, which is a standard result and can be found in, for instance, [54].

Theorem 14.2 *For all x in $\mathbb{R}$,*

$$(Ae_k)(x) = (2k)^{1/2}e_{k-1}(x), \quad k = 1, 2, \ldots,$$

and

$$(A^*e_k)(x) = (2k+2)^{1/2}e_{k+1}(x), \quad k = 0, 1, 2, \ldots.$$

Remark 14.3 The differential operator A lowers the order of the Hermite function from k to $k-1$, and the differential operator A^* lifts the order from k to $k+1$. Hence we call A and A^* the annihilation operator and the creation operator, respectively, for the Hermite functions e_k, $k = 0, 1, 2, \ldots$, on $\mathbb{R}$.

An immediate consequence of Theorem 14.2 is the following theorem.

Theorem 14.4 $He_k = (2k+1)e_k, \quad k = 0, 1, 2, \ldots.$

The proofs of Theorems 14.2 and 14.4 are left as exercises.

Remark 14.5 Theorem 14.4 says that for $k = 0, 1, 2, \ldots$, the positive integer $2k+1$ is an eigenvalue of the Hermite operator H and the Hermite function e_k on $\mathbb{R}$ is an eigenfunction of H corresponding to the eigenvalue $2k+1$. The differential operator H is the Schrödinger operator or the so-called Hamiltonian of a simple harmonic oscillator in quantum mechanics. The eigenvalues $2k+1$, $k = 0, 1, 2, \ldots$, are the results we get when we measure the energy of the oscillator. For $k = 0, 1, 2, \ldots$, the Hermite function e_k is the state of the oscillator in which the energy $2k+1$ is measured.

Remark 14.6 For $k = 0, 1, 2, \ldots$, can we find the Hermite function e_k by solving the Schrödinger equation

$$y'' + (2k + 1 - x^2)y = 0$$

directly using power series? If we try $y(x) = \sum_{n=0}^{\infty} a_n x^n$ as a solution, then we get a three-term recurrence relation for the coefficients of the power series. A three-term recurrence relation is difficult to handle. This is why we avoid solving the Schrödinger equation using power series.

The following formula, known as Mehler's formula, will be used in the construction of the heat kernel of the Hermite operator. A proof can be found in the book [54].

Theorem 14.7 *For all x and y in $\mathbb{R}$, and all w in $\mathbb{C}$ with $|w| < 1$,*

$$\sum_{k=0}^{\infty} e_k(x) e_k(y) w^k = \frac{1}{\sqrt{\pi}} (1 - w^2)^{-1/2} e^{-\frac{1}{2} \frac{1+w^2}{1-w^2}(x^2+y^2) + \frac{2w}{1-w^2}xy},$$

and the series is uniformly and absolutely convergent on all compact subsets of the disk $\{w \in \mathbb{C} : |w| < 1\}$.

Remark 14.8 To be specific, we use the principal branch of $(1 - w^2)^{-1/2}$, i.e.,

$$(1 - w^2)^{-1/2} = e^{-\frac{1}{2} \mathrm{Log}\,(1-w^2)},$$

where

$$\mathrm{Log}\,\zeta = \ln |\zeta| + i \mathrm{Arg}\,\zeta, \quad -\pi < \mathrm{Arg}\,\zeta < \pi.$$

Thus, $(1 - w^2)^{-1/2}$ is holomorphic on $\{w \in \mathbb{C} : |w| < 1\}$ and for any w in $\mathbb{R}$ with $|w| < 1$, we get

$$(1 - w^2)^{-\frac{1}{2}} = e^{-\frac{1}{2} \ln |1 - w^2|} > 0.$$

More details can be found in the book [56].

We can now construct the heat kernel of the Hermite operator $-\Delta + |x|^2$ on $\mathbb{R}^n$.

Theorem 14.9 *For every multi-index α,*

$$(-\Delta + |x|^2) e_\alpha = (2|\alpha| + 1) e_\alpha,$$

where

$$e_\alpha(x) = \prod_{j=1}^{n} e_{\alpha_j}(x_j), \quad x \in \mathbb{R}^n,$$

which is a tensor product of the one-dimensional Hermite functions.

Proof By Theorem 14.4, we get for all x in $\mathbb{R}^n$,

$$
\begin{aligned}
((-\Delta + |x|^2)e_\alpha)(x) &= \sum_{j=1}^n \left(-\frac{\partial^2}{\partial x_j^2} + x_j^2\right) e_{\alpha_1}(x_1) \cdots e_{\alpha_n}(x_n) \\
&= \sum_{j=1}^n (2\alpha_j + 1)e_{\alpha_1}(x_1) \cdots e_{\alpha_n}(x_n) \\
&= \sum_{j=1}^n (2\alpha_j + 1)e_\alpha(x) = (2|\alpha| + 1)e_\alpha(x).
\end{aligned}
$$

$\square$

Let $\mathbb{N}_0^n$ be the set of all multi-indices. Then the following fact is left as an exercise.

Theorem 14.10 $\{e_\alpha : \alpha \in \mathbb{N}_0^n\}$ *is an orthonormal basis for* $L^2(\mathbb{R}^n)$.

We can now give a formula for the solution of the initial value problem for the heat equation given by

$$
\begin{cases}
\frac{\partial u}{\partial t}(x,t) = ((\Delta - |x|^2)u)(x,t), & x \in \mathbb{R}^n, t > 0, \\
u(x,0) = f(x), & x \in \mathbb{R}^n,
\end{cases}
\tag{14.2}
$$

where f is a Schwartz function on $\mathbb{R}^n$. It follows from series expansions in Hilbert spaces, Theorems 14.9, and 14.10 that a solution u is given by

$$
u(x,t) = (e^{-(-\Delta + |x|^2)t}f)(x) = \sum_{\alpha \in \mathbb{N}_0^n} e^{-(2|\alpha|+1)t}(f, e_\alpha)e_\alpha(x), \quad x \in \mathbb{R}^n, t > 0,
$$

where

$$
(f, e_\alpha) = \int_{\mathbb{R}^n} f(y)e_\alpha(y)\, dy, \quad \alpha \in \mathbb{N}_0^n.
$$

We are now ready to prove the following theorem, which gives a formula for the heat kernel.

Theorem 14.11 *Let* $f \in S$. *Then for all* x *in* $\mathbb{R}^n$ *and* $t > 0$,

$$
(e^{-(-\Delta + |x|^2)t}f)(x) = \int_{\mathbb{R}^n} H_t(x,y)f(y)\, dy,
$$

where

$$
H_t(x,y) = \frac{1}{\pi^n}e^{-t}(1 - e^{-4t})^{-n/2}e^{-\frac{1}{2}(|x|^2+|y|^2)\coth(2t))+(x\cdot y)\operatorname{csch}(2t)}, \quad t > 0.
$$

We call the function H_t, $t > 0$, on $\mathbb{R}^n \times \mathbb{R}^n$ the heat kernel of the Hermite operator. The operator $e^{-(-\Delta + |x|^2)t}$, $t > 0$, is called the heat semigroup of the Hermite operator.

Proof of Theorem 14.11 For all x in $\mathbb{R}^n$ and $t > 0$, we get by means of the basic series expansions in Hilbert spaces

$$(e^{-(-\Delta+|x|^2)t}f)(x) = \sum_{\alpha\in\mathbb{N}_0^n} e^{-(2|\alpha|+1)t}(f,e_\alpha)e_\alpha(x)$$

$$= \int_{\mathbb{R}^n}\left(\sum_{\alpha\in\mathbb{N}_0^n} e^{-(2|\alpha|+1)t}e_\alpha(x)e_\alpha(y)\right) f(y)\,dy.$$

Using Mehler's formula, we get

$$\sum_{\alpha\in\mathbb{N}_0^n} e^{-(2|\alpha|+1)t}e_\alpha(x)e_\alpha(y)$$

$$= e^{-t}\sum_{\alpha_1=0}^{\infty}\cdots\sum_{\alpha_n=0}^{\infty}[e^{-2\alpha_1 t}e_{\alpha_1}(x_1)e_{\alpha_1}(y_1)]\cdots[e^{-2\alpha_n t}e_{\alpha_n}(x_n)e_{\alpha_n}(y_n)]$$

$$= e^{-t}\frac{1}{\pi^n}(1-e^{-4t})^{-n/2}\prod_{j=1}^{n} e^{-\frac{1}{2}\frac{1+e^{-4t}}{1-e^{-4t}}(x_j^2+y_j^2)+\frac{2e^{-2t}}{1-e^{-4t}}x_j y_j}$$

$$= H_t(x,y),$$

for all x and y in $\mathbb{R}^n$ and $t > 0$, where

$$H_t(x,y) = \frac{1}{\pi^n}e^{-t}(1-e^{-4t})^{-n/2}e^{-\frac{1}{2}(|x|^2+|y|^2)\coth(2t)+(x\cdot y)\operatorname{csch}(2t)}.$$

Therefore for all x and y in $\mathbb{R}^n$ and $t > 0$,

$$(e^{-(-\Delta+|x|^2)t}f)(x) = \int_{\mathbb{R}^n} H_t(x,y)f(y)\,dy,$$

as asserted. $\qquad\square$

We end this chapter with the trace of the heat semigroup for the Hermite operator, which is the sum of the eigenvalues of the heat semigroup.

Theorem 14.12 $\operatorname{tr}(e^{-(-\Delta+|x|^2)t}) = \frac{e^{(n-1)t}}{2^n\sinh^n t}$, $\quad t > 0$.

Proof For $t > 0$,

$$\operatorname{tr}(e^{-(-\Delta+|x|^2)}) = \sum_{\alpha\in\mathbb{N}_0^n} e^{-(2|\alpha|+1)t} = e^{-t}\sum_{\alpha\in\mathbb{N}_0^n} e^{-2|\alpha|t}$$

$$= e^{-t}\left(\sum_{k=0}^{\infty} e^{-2kt}\right)^n = e^{-t}\left(\frac{1}{1-e^{-2t}}\right)^n$$

$$= e^{-t}\left(\frac{e^t}{e^t - e^{-t}}\right)^n = e^{-t}\frac{e^{nt}}{(2\sinh t)^n}$$

$$= \frac{e^{(n-1)t}}{2^n\sinh^n t}.$$

$\qquad\square$

Remark 14.13 Among the most useful formulas in this chapter is the Mehler formula for Hermite functions, which is Theorem 23.1 in the book [54]. This formula is used in Chapter 22 to give another version of Mehler's formula for Fourier–Wigner transforms of Hermite functions.

Historical Notes

Charles Hermite (1822–1901) was an eminent French mathematician of the 19th century. He was a professor at the Sorbonne. The Hermite functions are important not only as quantum states of a simple harmonic oscillator, but also for the development of Fourier analysis.

Harald Cramér (1893–1985) was a Swedish mathematician, actuary, and statistician. He earned his doctorate from Stockholm University under the direction of Marcel Riesz.

Gustav Ferdinand Mehler (1835–1895) was a German mathematician. Other than the Mehler formula in this chapter, his contributions in mathematical analysis include formulas in special functions and integral transforms.

Exercises

1. Prove that the Hermite polynomial $H_k(x)$ is a solution of the differential equation
$$y'' - 2xy' + 2ky = 0$$
for $k = 0, 1, 2, \ldots$. (The differential equation $y'' - 2xy' + \lambda y = 0$, where λ is a real number is known as the Hermite equation.)

2. Use power series to find the polynomial solution of the differential equation
$$y'' - 2xy' + 2ky = 0$$
for $k = 0, 1, 2, \ldots$.

3. For $k = 0, 1, 2, \ldots$, let P_k be the polynomial solution of
$$y'' - 2xy' + 2ky = 0$$
for which the coefficient of x^k is 2^k. Compute P_0, P_1, P_2, and P_3. Compute also $(-1)^k e^{x^2} \left(\frac{d}{dx}\right)^k (e^{-x^2})$ for all x in $\mathbb{R}$ and $k = 0, 1, 2, 3$. Make a conjecture.

4. Let w be the function on $\mathbb{R}$ defined by
$$w(x) = e^{x^2/2} y(x), \quad x \in \mathbb{R}.$$
Prove that the Hermite equation
$$y'' - 2xy' + \lambda y = 0,$$
where λ is a real number, is the same as the differential equation
$$w'' + (\lambda - x^2)w = 0.$$

5. Compute the three-term recurrence relation for a power series solution

$$y(x) = \sum_{n=0}^{\infty} a_n x^n$$

of the Schrödinger equation

$$y'' + (\lambda - x^2)y = 0,$$

where λ is a real number.

6. Let $A = \frac{d}{dx} + x$ and $A^* = -\frac{d}{dx} + x$. Prove that

$$\int_{-\infty}^{\infty} (Au)(x)\overline{v(x)}\, dx = \int_{-\infty}^{\infty} u(x)\overline{(A^*v)(x)}\, dx$$

for all functions u and v in $\mathcal{S}$.

7. Let

$$H = -\frac{1}{2}(AA^* + A^*A),$$

where A and A^* are as in the preceding exercise. Prove that

$$H = -\frac{d^2}{dx^2} + x^2.$$

8. Prove Theorem 14.2.

9. Prove Theorem 14.4.

10. Prove that $\{e_\alpha : \alpha \in \mathbb{N}_0^n\}$ is an orthonormal basis for $L^2(\mathbb{R}^n)$. (Hint: You may use the fact that $\{e_k : k = 0, 1, 2, \dots\}$ is an orthonormal basis for $L^2(\mathbb{R})$ and also the fact that the set of all finite linear combinations of tensor products of functions in $L^2(\mathbb{R})$ is dense in $L^2(\mathbb{R}^n)$.)

11. Let $f \in \mathcal{S}$ and let u be the function on $\mathbb{R}^n \times (0, \infty)$ defined by

$$u(x, t) = \int_{\mathbb{R}^n} H_t(x, y)\, f(y)\, dy, \quad x \in \mathbb{R}^n,\ t > 0.$$

Prove that u is a solution of (14.2) for which $f \in \mathcal{S}$. What happens to $u(\cdot, t)$ as $t \to \infty$?

12. Prove that if $f \in L^2(\mathbb{R}^n)$, then the function u on $\mathbb{R}^n \times (0, \infty)$ defined by

$$u(x, t) = \int_{\mathbb{R}^n} H_t(x, y)f(y)\, dy, \quad x \in \mathbb{R}^n,\ t > 0,$$

is a solution of (14.2) for which $f \in L^2(\mathbb{R}^n)$. What happens to $u(\cdot, t)$ as $t \to \infty$?

Chapter 15

The Green Function of the Hermite Operator

We give in this chapter a solution u of the Poisson equation

$$(-\Delta + |x|^2)u = f \qquad (15.1)$$

for the Hermite operator, where f is a Schwartz function on $\mathbb{R}^n$. By Theorem 14.4 and the series expansions in Hilbert spaces, it is given by

$$
\begin{aligned}
u(x) &= \sum_{\alpha \in \mathbb{N}_0^n} \frac{1}{2|\alpha|+1}(f, e_\alpha)e_\alpha(x) \\
&= \sum_{\alpha \in \mathbb{N}_0^n} \frac{1}{2|\alpha|+1}\left(\int_{\mathbb{R}^n} f(y)e_\alpha(y)\,dy\right)e_\alpha(x) \\
&= \int_{\mathbb{R}^n}\left(\sum_{\alpha \in \mathbb{N}_0^n} \frac{1}{2|\alpha|+1}e_\alpha(x)e_\alpha(y)\right)f(y)\,dy, \quad x \in \mathbb{R}^n.
\end{aligned}
$$

Thus, the Green function of the Hermite operator is the function G on $\mathbb{R}^n \times \mathbb{R}^n$ given by

$$G(x,y) = \sum_{\alpha \in \mathbb{N}_0^n} \frac{1}{2|\alpha|+1}e_\alpha(x)e_\alpha(y) \qquad (15.2)$$

for all x and y in $\mathbb{R}^n$. In fact,

$$
\begin{aligned}
G(x,y) &= \int_0^\infty H_t(x,y)\,dt \\
&= \frac{1}{\pi^n}\int_0^\infty e^{-t}(1-e^{-4t})^{-n/2}e^{-\frac{1}{2}(|x|^2+|y|^2)\coth(2t)+(x\cdot y)\operatorname{csch}(2t)}\,dt.
\end{aligned}
$$

Remark 15.1 The formula for the Green function in integral form is not as tractable as one expects. It is the series (15.2) that turns out to be more useful for the Hermite operator.

We now give a uniqueness result on the solution u of (15.1).

DOI: 10.1201/9781003206781-15

Theorem 15.2 *Let $u \in L^2(\mathbb{R}^n)$ be a solution of the Hermite equation*

$$(-\Delta + |x|^2)u = 0$$

on $\mathbb{R}^n$. Then $u = 0$.

Proof Since

$$u = \sum_{\alpha \in \mathbb{N}_0^n} (u, e_\alpha) e_\alpha,$$

it follows that

$$\sum_{\alpha \in \mathbb{N}_0^n} (u, e_\alpha)(2|\alpha| + 1) e_\alpha = 0.$$

So, by Parseval's identity, we get

$$(2|\alpha| + 1)^2 |(u, e_\alpha)|^2 = 0, \quad \alpha \in \mathbb{N}_0^n.$$

Therefore

$$(u, e_\alpha) = 0, \quad \alpha \in \mathbb{N}_0^n.$$

Since $\{e_\alpha : \alpha \in \mathbb{N}_0^n\}$ is an orthonormal basis for $L^2(\mathbb{R}^n)$, it follows that $u = 0$, as asserted. $\qquad\square$

We can now consider the partial differential equation (15.1) when $f \in \mathcal{S}'$. To solve for u, we use the following theorem.

Theorem 15.3 *Let $f \in \mathcal{S}'$. Then there exists a β in $\mathbb{N}_0^n$ such that*

$$|f(e_\alpha)| \le C(|\alpha| + 1)^{|\beta|}, \quad \alpha \in \mathbb{N}_0^n.$$

Moreover,

$$f = \sum_{\alpha \in \mathbb{N}_0^n} f(e_\alpha) e_\alpha,$$

where the convergence of the series is in $\mathcal{S}'$.

The proof of Theorem 15.3 entails several interesting results about the Schwartz space and tempered distributions, which are largely based on the results in [34, 40].

Let us recall that the Schwartz space $\mathcal{S}$ is topologized by the norms $\| \ \|_{\alpha,\beta,\infty}$ for all α and β in $\mathbb{N}_0^n$, where

$$\|\varphi\|_{\alpha,\beta,\infty} = \sup_{x \in \mathbb{R}^n} |x^\alpha (\partial^\beta \varphi)(x)|, \quad \varphi \in \mathcal{S}.$$

Let α and β be in $\mathbb{N}_0^n$. Then for all φ in $\mathcal{S}$, we define $\|\varphi\|_{\alpha,\beta,2}$ by

$$\|\varphi\|_{\alpha,\beta,2}^2 = \int_{\mathbb{R}^n} |x^\alpha (\partial^\beta \varphi)(x)|^2 dx.$$

Lemma 15.4 $\{\|\ \|_{\alpha,\beta,\infty} : \alpha, \beta \in \mathbb{N}_0^n\}$ *and* $\{\|\ \|_{\alpha,\beta,2} : \alpha, \beta \in \mathbb{N}_0^n\}$ *are equivalent families of norms in $\mathcal{S}$. More precisely, for all α and β in $\mathbb{N}_0^n$, we can find a positive constant C_1, multi-indices γ_j and δ_j, $j = 1, 2, \ldots, N$, such that*

$$\|\varphi\|_{\alpha,\beta,\infty} \leq C_1(\|\varphi\|_{\gamma_1,\delta_1,2} + \|\varphi\|_{\gamma_2,\delta_2,2} + \cdots + \|\varphi\|_{\gamma_N,\delta_N,2}). \quad \varphi \in \mathcal{S}.$$

and for all γ and δ in $\mathbb{N}_0^n$, we can find a positive constant C_2, multi-indices α_j and β_j, $j = 1, 2, \ldots, M$, such that

$$\|\varphi\|_{\gamma,\delta,2} \leq C_2(\|\varphi\|_{\alpha_1,\beta_1,\infty} + \|\varphi\|_{\alpha_2,\beta_2,\infty} + \cdots + \|\varphi\|_{\alpha_M,\beta_M,\infty}), \quad \varphi \in \mathcal{S}.$$

Proof Let $f \in \mathcal{S}$. Then

$$
\begin{aligned}
\|f\|_2 &= \left(\int_{\mathbb{R}^n} |f(x)|^2 dx \right)^{1/2} \\
&= \left(\int_{\mathbb{R}^n} (1 + |x|^2)^{-2n}(1 + |x|^2)^{2n}|f(x)|^2 dx \right)^{1/2} \\
&\leq C\|(1 + |x|^2)^n f\|_\infty,
\end{aligned}
$$

where

$$C = \left(\int_{\mathbb{R}^n} (1 + |x|^2)^{-2n} dx \right)^{1/2}. \tag{15.3}$$

So, for all φ in $\mathcal{S}$,

$$\|\varphi\|_{\alpha,\beta,2} = \|x^\alpha D^\beta \varphi\|_2 \leq C\|(1 + |x|^2)^n x^\alpha D^\beta \varphi\|_\infty.$$

Let P be the polynomial in $\mathbb{R}^n$ such that

$$P(x) = (1 + |x|^2)^n x^\alpha, \quad x \in \mathbb{R}^n.$$

Writing

$$P(x) = \sum_{|\gamma| \leq m} a_\gamma x^\gamma, \quad x \in \mathbb{R}^n,$$

we get

$$\|\varphi\|_{\alpha,\beta,2} \leq C \sum_{|\gamma| \leq m} |a_\gamma| \|\varphi\|_{\alpha+\gamma,\beta,\infty}.$$

Next, let $\mathbf{1}$ be the multi-index $(1, 1, \ldots, 1)$. Then by the fundamental theorem of calculus, we get for all differentiable functions f on $\mathbb{R}^n$,

$$f(x) = \int_{-\infty}^{x_n} \cdots \int_{-\infty}^{x_1} (\partial^{\mathbf{1}} f)(t_1, \ldots, t_n) \, dt_1 \cdots dt_n, \quad x \in \mathbb{R}^n.$$

Therefore by the Schwarz inequality,

$$
\begin{aligned}
\|f\|_\infty &\leq \int_{-\infty}^{x_n} \cdots \int_{-\infty}^{x_1} |(\partial^{\mathbf{1}} f)(t_1, \ldots, t_n)| \, dt_1 \cdots dt_n \leq \|\partial^{\mathbf{1}} f\|_1 \\
&\leq C\|(1 + |x|^2)^n (\partial^{\mathbf{1}} f)\|_2,
\end{aligned}
$$

where C is given by (15.3). Let $\varphi \in \mathcal{S}$. Then for all multi-indices α and β, we use the Leibniz formula in Proposition 1.1 and (1.9)

$$\partial^1 (x^\alpha \partial^\beta \varphi) = \sum_{\gamma \leq 1} \binom{1}{\gamma} \binom{\alpha}{\gamma} x^{\alpha - \gamma} \partial^{1 - \gamma + \beta} f,$$

where $\binom{\alpha}{\gamma} = 0$ if γ is not $\leq \alpha$. Thus,

$$\|\varphi\|_{\alpha, \beta, \infty} \leq C \left\| \sum_{\gamma \leq 1} \binom{1}{\gamma} \binom{\alpha}{\gamma} (1 + |x|^2)^n x^{\alpha - \gamma} \partial^{1 - \gamma + \beta} f \right\|_2.$$

Writing

$$(1 + |x|^2)^n x^{\alpha - \gamma} = \sum_{|\delta| \leq m_{\alpha, \gamma}} b_\delta x^\delta, \quad x \in \mathbb{R}^n,$$

we obtain

$$\|\varphi\|_{\alpha, \beta, \infty} \leq C \sum_{\gamma \leq 1} \binom{1}{\gamma} \binom{\alpha}{\gamma} \sum_{|\delta| \leq m_{\alpha, \delta}} |b_\delta| \|\varphi\|_{\delta, 1 - \gamma - \beta, 2}.$$

This completes the proof.   □

For $j = 1, 2, \ldots, n$, let A_j and A_j^* be the differential operators defined by

$$A_j = \frac{\partial}{\partial x_j} + x_j$$

and

$$A_j^* = -\frac{\partial}{\partial x_j} + x_j,$$

and let P_j be the operator defined by

$$P_j = \frac{1}{2} A_j^* A_j.$$

Then for every multi-index β,

$$(P^\beta e_\alpha)(x) = \prod_{j=1}^n P_j^{\beta_k} \left(\prod_{k=1}^n e_{\alpha_k}(x_k) \right) = \alpha^\beta e_\alpha(x), \quad x \in \mathbb{R}^n. \tag{15.4}$$

For every γ in $\mathbb{N}_0^n$, we define the norm $\|\| \ \|\|_\gamma$ in $\mathcal{S}$ by

$$\|\|\varphi\|\|_\gamma = \|Q^\gamma \varphi\|_2$$

for all $\varphi \in \mathcal{S}$, where

$$Q^\gamma = Q_1^{\gamma_1} Q_2^{\gamma_2} \cdots Q_n^{\gamma_n}$$

and

$$Q_j = P_j + 1, \quad j = 1, 2, \ldots, n.$$

Lemma 15.5 $\{\|\| \; \|\|_\gamma : \gamma \in \mathbb{N}_0^n\}$ *is a directed family of norms in* $\mathcal{S}$ *equivalent to the family* $\{\| \; \|_{\alpha,\beta,2} : \alpha, \beta \in \mathbb{N}_0^n\}$. *Here, a directed family of norms means that for all multi-indices* γ *and* κ *in* $\mathbb{N}_0^n$, *we can find a positive constant* C *and a multi-index* ω *such that*

$$\|\|\varphi\|\|_\gamma + \|\|\varphi\|\|_\kappa \leq C\|\|\varphi\|\|_\omega$$

for all φ *in* $\mathcal{S}$.

Proof Let $\varphi \in \mathcal{S}$. Then for every multi-index γ,

$$
\begin{aligned}
\|\|\varphi\|\|_\gamma^2 &= \|Q^\gamma \varphi\|_2^2 \\
&= \|(P_1 + 1)^{\gamma_1} \cdots (P_n + 1)^{\gamma_n} \varphi\|_2^2 \\
&= \left\| \sum_{\delta_1 \leq \gamma_1} \cdots \sum_{\delta_n \leq \gamma_n} \binom{\gamma_1}{\delta_1} \cdots \binom{\gamma_n}{\delta_n} P^\delta \varphi \right\|_2^2 \\
&= \left\| \sum_{\delta \leq \gamma} \binom{\gamma}{\delta} P^\delta \varphi \right\|_2^2 .
\end{aligned}
\tag{15.5}
$$

Using (15.5), Parseval's identity, and (15.4),

$$
\begin{aligned}
\|\|\varphi\|\|_\gamma^2 &= \left\| \sum_{\delta \leq \gamma} \binom{\gamma}{\delta} P^\delta \sum_{\alpha \in \mathbb{N}_0^n} (\varphi, e_\alpha) e_\alpha \right\|_2^2 \\
&= \left\| \sum_{\delta \leq \gamma} \binom{\gamma}{\delta} \sum_{\alpha \in \mathbb{N}_0^n} (\varphi, e_\alpha) \alpha^\delta e_\alpha \right\|_2^2 \\
&= \left\| \sum_{\alpha \in \mathbb{N}_0^n} \left\{ (\varphi, e_\alpha) \sum_{\delta \leq \gamma} \binom{\gamma}{\delta} \alpha^\delta \right\} e_\alpha \right\|_2^2 .
\end{aligned}
\tag{15.6}
$$

So, applying Parseval's identity to (15.6), we get

$$\|\|\varphi\|\|_\gamma^2 = \sum_{\alpha \in \mathbb{N}_0^n} |(\varphi, e_\alpha)|^2 \left| \sum_{\delta \leq \gamma} \binom{\gamma}{\delta} \alpha^\delta \right|^2 .$$

Similarly, for every multi-index κ, we get

$$\|\|\varphi\|\|_\kappa^2 = \sum_{\alpha \in \mathbb{N}_0^n} |(\varphi, e_\alpha)|^2 \left| \sum_{\delta \leq \kappa} \binom{\kappa}{\delta} \alpha^\delta \right|^2 .$$

and we can choose ω to be $\gamma + \kappa$ to conclude that $\{\||\ \||_\gamma : \gamma \in \mathbb{N}_0^n\}$ is a directed family. To prove that the two families of norms are equivalent, we first establish the following inequality, which says that for all functions φ in $\mathcal{S}$,

$$\|A^{(1)}A^{(2)}\cdots A^{(m)}\varphi\|_2 \le \|(2P+2m\mathbf{1})^\gamma \varphi\|_2, \qquad (15.7)$$

where $A^{(1)}, A^{(2)}, \ldots, A^{(n)}$ are arbitrary but fixed operators from A_j and A_j^* for $j = 1, 2, \ldots, n$, γ is a multi-index with $|\gamma| = m$. Indeed, by Parseval's identity and Theorem 14.2,

$$\|A^{(1)}A^{(2)}\cdots A^{(m)}\varphi\|_2^2$$

$$= \left\|\sum_{\alpha\in\mathbb{N}_0^n}(\varphi,e_\alpha)A^{(1)}A^{(2)}\cdots A^{(m)}e_\alpha\right\|_2^2$$

$$= \left\|\sum_{\alpha\in\mathbb{N}_0^n}(\varphi,e_\alpha)e_{\alpha_1\pm\gamma_1,\alpha_2\pm\gamma_2,\ldots,\alpha_n\pm\gamma_n}\varepsilon_1^{\gamma_1}\varepsilon_2^{\gamma_2}\cdots\varepsilon_n^{\gamma_n}\right\|_2^2,$$

where $\gamma \le \alpha$,

$$0 \le \gamma_j \le m, \quad j = 1, 2, \ldots, n,$$

and

$$0 \le \varepsilon_j \le \sqrt{2\alpha_j + 2m}, \quad j = 1, 2, \ldots, n.$$

Using Bessel's inequality and Parseval's identity for Hilbert spaces, we get

$$\|A^{(1)}A^{(2)}\cdots A^{(m)}\varphi\|_2^2$$

$$\le \left\|\sum_{\alpha\in\mathbb{N}_0^n}(\varphi,e_\alpha)e_{\alpha\pm\gamma}\prod_{j=1}^n \varphi_j^{\gamma_j}\right\|_2^2$$

$$\le \sum_{\alpha\in\mathbb{N}_0^n}|(\varphi,e_\alpha)|^2\prod_{j=1}^n \varepsilon_j^{2\gamma_j}$$

$$\le \sum_{\alpha\in\mathbb{N}_0^n}|(\varphi,e_\alpha)|^2\prod_{j=1}^n (2\alpha_j+2m)^{\gamma_j}$$

$$= \sum_{|\beta\mp\gamma|\ge0}|(\varphi,e_{\beta\mp\gamma})|^2\prod_{j=1}^n (2(\beta_j\mp\gamma_j+2m))^{\gamma_j}$$

$$\le \sum_{\alpha\in\mathbb{N}_0^n}|(\varphi,e_\alpha)|^2\prod_{j=1}^n (2\alpha_j+2m)^{\gamma_j},$$

where $\alpha \pm \gamma$ refers to the multi-index in which the jth entry is $\alpha_j + \gamma_j$ or $\alpha_j - \gamma_j$ for $j = 1, 2, \ldots, n$, and the same meaning applies to $\beta \mp \gamma$. Since

$$2P_j = -\frac{\partial^2}{\partial x_j^2} + x_j^2 - 1, \quad j = 1, 2, \ldots, n,$$

it follows that

$$(2P + 2m\mathbf{1})^\gamma e_\alpha = \prod_{j=1}^{n} (2\alpha_j + 2m)^{\gamma_j} e_\alpha.$$

Thus, by Parseval's identity,

$$\|(2P + 2m\mathbf{1})^\gamma \varphi\|_2^2$$

$$= \left\| \sum_{\alpha \in \mathbb{N}_0^n} (\varphi, e_\alpha)(2P + 2m\mathbf{1})^\gamma e_\alpha \right\|_2^2$$

$$= \left\| \sum_{\alpha \in \mathbb{N}_0^n} (\varphi, e_\alpha) e_\alpha \prod_{j=1}^{n} (2\alpha_j + 2m)^{\gamma_j} \right\|_2^2$$

$$= \sum_{\alpha \in \mathbb{N}_0^n} |(\varphi, e_\alpha)|^2 \prod_{j=1}^{n} (2\alpha_j + 2m)^{2\gamma_j},$$

and (15.7) is proved. Let γ be a multi-index. Then for all φ in $\mathcal{S}$, we can use (15.5) to obtain

$$|||\varphi|||_\gamma = \left\| \sum_{\delta \leq \gamma} \binom{\gamma}{\delta} P^\delta \varphi \right\|_2 = \left\| \sum_{\delta \leq \gamma} 2^{-|\delta|} \binom{\gamma}{\delta} \prod_{j=1}^{n} (A_j^* A_j)^{\delta_j} \varphi \right\|_2. \tag{15.8}$$

Carrying out the differentiations and rearranging terms so that multiplication by a monomial appears before differentiations in each summand in (15.8), we can find a positive constant C_1, and multi-indices $\alpha_1, \alpha_2, \ldots, \alpha_N$ and $\beta_1, \beta_2, \ldots, \beta_N$ such that

$$|||\varphi|||_\gamma \leq C(\|\varphi\|_{\alpha_1, \beta_1, 2} + \|\varphi\|_{\alpha_2, \beta_2, 2} + \cdots + \|\varphi\|_{\alpha_N, \beta_N, 2}).$$

On the other hand, let α and β be multi-indices. Then

$$x^\alpha \partial^\beta \varphi = \prod_{j=1}^{n} x_j^{\alpha_j} \prod_{k=1}^{n} \partial_k^{\beta_k} \varphi$$

$$= 2^{-|\alpha+\beta|} \prod_{j=1}^{n} (A_j + A_j^*)^{\alpha_j} \prod_{k=1}^{n} (A_k - A_k^*)^{\beta_j} \varphi,$$

which is a finite linear combination of terms of the form

$$A^{(1)} A^{(2)} \cdots A^{(m)}$$

in (15.7). So, by (15.7), we can find a positive constant C_2 and multi-indices

$\gamma^{(1)}, \gamma^{(2)}, \dots, \gamma^{(M)}$ such that

$$
\begin{aligned}
\|\varphi\|_{\alpha,\beta,2} &\leq C_2 \sum_{j=1}^{M} \|(2P + 2|\gamma^{(j)}|\mathbf{1})^{\gamma^{(j)}} \varphi\|_2 \\
&\leq C_2 \sum_{j=1}^{M} (2|\gamma^{(j)}|)^{|\gamma^{(j)}|} \|(P+1)^{\gamma^{(j)}} \varphi\|_2 \\
&= C_2 \sum_{j=1}^{M} (2|\gamma^{(j)}|)^{|\gamma^{(j)}|} \|\varphi\|_{\gamma^{(j)}},
\end{aligned}
$$

and the proof is complete. $\qquad\qquad\qquad\qquad\qquad\qquad\qquad\square$

Let $f \in \mathcal{S}$. Then

$$
f = \sum_{\alpha \in \mathbb{N}_0^n} (f, e_\alpha) e_\alpha,
$$

where the convergence is in $L^2(\mathbb{R}^n)$. So, for all multi-indices β,

$$
Q^\beta f = \sum_{\alpha \in \mathbb{N}_0^n} (Q^\beta f, e_\alpha) e_\alpha = \sum_{\alpha \in \mathbb{N}_0^n} (f, Q^\beta e_\alpha) e_\alpha = \sum_{\alpha \in \mathbb{N}_0^n} (1+\alpha)^\beta (f, e_\alpha) e_\alpha.
$$

Thus,

$$
\sum_{\alpha \in \mathbb{N}_0^n} |(1+\alpha)^\beta f(e_\alpha)|^2 < \infty
$$

and hence

$$
\sup_{\alpha \in \mathbb{N}_0^n} |(1+\alpha)^\beta f(e_\alpha)| < \infty \qquad\qquad (15.9)
$$

for all multi-indices β.

Proof of Theorem 15.3 We first note that for all φ in $\mathcal{S}$,

$$
\sum_{\alpha \in \mathbb{N}_0^n} |f(e_\alpha) e_\alpha(\varphi)| < \infty.
$$

Indeed, by the continuity of f on the Schwartz space with respect to the family $\{\|| \; \||_\gamma : \gamma \in \mathbb{N}_0^n\}$, we can find a positive constant C and a multi-index γ such that

$$
|f(e_\alpha)| \leq C \|| e_\alpha \||_\gamma = C \|Q^\gamma e_\alpha\|_2
$$

for all multi-indices α. Since

$$
Q^\gamma e_\alpha = (1+\alpha)^\gamma e_\alpha
$$

for all multi-indices α, it follows that

$$
|f(e_\alpha)| \leq C|(1+\alpha)^\gamma|, \quad \alpha \in \mathbb{N}_0^n. \qquad\qquad (15.10)
$$

So, by (15.9) and (15.10), there exists a positive constant C such that

$$\sum_{\alpha \in \mathbb{N}_0^n} |f(e_\alpha)e_\alpha(\varphi)| \le C \sum_{\alpha \in \mathbb{N}_0^n} |(1+\alpha)^\gamma| |(1+\alpha)^\beta|^{-1} < \infty$$

if we choose β large enough. Therefore there exists a tempered distribution g such that

$$g = \sum_{\alpha \in \mathbb{N}_0^n} f(e_\alpha)e_\alpha$$

in $\mathcal{S}'$. So, for all β in $\mathbb{N}_0^n$,

$$g(e_\beta) = \sum_{\alpha \in \mathbb{N}_0^n} f(e_\alpha)e_\alpha(e_\beta) = f(e_\beta).$$

Thus, $g = f$ and

$$f = \sum_{\alpha \in \mathbb{N}_0^n} f(e_\alpha)e_\alpha,$$

as claimed. □

To solve the partial differential differential equation (15.1) for a tempered distribution f on $\mathbb{R}^n$, we let u be the tempered distribution defined by

$$u = \sum_{\alpha \in \mathbb{N}_0^n} \frac{1}{2|\alpha| + 1} f(e_\alpha)e_\alpha, \tag{15.11}$$

where the convergence is understood in the sense of tempered distributions. Then for all functions φ in $\mathcal{S}$,

$$
\begin{aligned}
&((-\Delta + |x|^2)u)(\varphi) \\
={}& u((-\Delta + |x|^2)\varphi)) = \sum_{\alpha \in \mathbb{N}_0^n} \frac{1}{2|\alpha| + 1} f(e_\alpha)e_\alpha((-\Delta + |x|^2)\varphi) \\
={}& \sum_{\alpha \in \mathbb{N}_0^n} \frac{1}{2|\alpha| + 1} f(e_\alpha)((-\Delta + |x|^2)e_k)(\varphi) \\
={}& \sum_{\alpha \in \mathbb{N}_0^n} \frac{1}{2|\alpha| + 1} f(e_\alpha)(2|\alpha| + 1)e_k(\varphi) \\
={}& \sum_{\alpha \in \mathbb{N}_0^n} f(e_\alpha)e_\alpha(\varphi) \\
={}& f(\varphi)
\end{aligned}
$$

and hence u is indeed a solution.

Historical Notes

Marc-Antoine Parseval (1755–1836) was a French mathematician best known for his identity for Fourier series. The Parseval identity has been developed to much more powerful results known as Plancherel's theorems.

Karl Hermann Amandus Schwarz (1843–1921), a German mathematician, has made important contributions in complex analysis. We just mention here the Schwarz lemma providing useful information on holomorphic functions from the unit disk D centered at the origin into D and the Schwarz problem, which is a Dirichlet problem for holomorphic functions on D. The Schwarz inequality used in this book is a versatile inequality in mathematics.

Exercises

1. Let $f \in \mathcal{S}$. Prove that the function u defined on $\mathbb{R}^n$ by

$$u(x) = \int_{\mathbb{R}^n} G(x, y) \, f(y) \, dy, \quad x \in \mathbb{R}^n,$$

is a solution of

$$((-\Delta + |x|^2)u)(x) = f(x), \quad x \in \mathbb{R}^n.$$

2. Let $s \in \mathbb{C}$ be such that $\operatorname{Re} s > 1$. Prove that the trace

$$\operatorname{tr}\left(\left(-\frac{d^2}{dx^2} + x^2\right)^{-s}\right)$$

of the inverse of $(-\frac{d^2}{dx^2} + x^2)^s$ is given by

$$\operatorname{tr}\left(\left(-\frac{d^2}{dx^2} + x^2\right)^{-s}\right) = (1 - 2^{-s})\zeta(s),$$

where ζ is the Riemann zeta-function defined by

$$\zeta(s) = \sum_{k=1}^{\infty} \frac{1}{k^s}, \quad \operatorname{Re} s > 1.$$

3. Compute the trace of the inverse of $(-\frac{d^2}{dx^2} + x^2)^2$.

Chapter 16

Global Regularity of the Hermite Operator

We begin this chapter with a result on the global hypoellipticity in the Schwartz space of the Hermite operator.

Theorem 16.1 *Let u be a tempered distribution on $\mathbb{R}^n$. Then*

$$(-\Delta + |x|^2)u \in \mathcal{S} \Rightarrow u \in \mathcal{S}.$$

The proof depends on the following lemma.

Lemma 16.2 *Let $\{a_\alpha : \alpha \in \mathbb{N}_0^n\}$ be such that*

$$\sup_{\alpha \in \mathbb{N}_0^n} \left(|a_\alpha| \, |\alpha|^m \right) < \infty \tag{16.1}$$

for all nonnegative integers m. Then there exists a unique function f in $\mathcal{S}$ such that

$$f = \sum_{\alpha \in \mathbb{N}_0^n} a_\alpha e_\alpha,$$

where the convergence of the series is in $\mathcal{S}$.

Proof For all positive integers N, let

$$f_N = \sum_{|\alpha| \leq N} a_\alpha e_\alpha.$$

Let γ be a multi-index. Then for all positive integers N and M with $M \geq N$,

DOI: 10.1201/9781003206781-16

we get by means of (15.4)

$$
\begin{aligned}
& \|f_N - f_M\|_\gamma^2 \\
= {} & \|Q^\gamma(f_N - f_M)\|_2^2 \\
= {} & \left\| \sum_{|\alpha| \le N} a_\alpha (1+\alpha)^\gamma e_\alpha - \sum_{|\alpha| \le M} a_\alpha (1+\alpha)^\gamma e_\alpha \right\|_2^2 \\
= {} & \left\| \sum_{N+1 \le |\alpha| \le M} a_\alpha (1+\alpha)^\gamma e_\alpha \right\|_2^2 \\
\le {} & \sum_{N+1 \le |\alpha| \le M} |a_\alpha|^2 |(1+\alpha)^\gamma|^2 \\
\le {} & \sum_{N+1 \le |\alpha| \le M} |\alpha|^{-2m} (n + |\alpha|)^{|\gamma|} \to 0
\end{aligned}
$$

as $N, M \to \infty$ provided that m is large enough. So, the sequence $\{f_N : N = 1, 2, \dots\}$ is Cauchy in $\mathcal{S}$ and converges to some function f in $\mathcal{S}$ as $N \to \infty$. Thus,

$$
f = \sum_{\alpha \in \mathbb{N}_0^n} a_\alpha e_\alpha,
$$

where the series is convergent in $\mathcal{S}$.   □

Proof of Theorem 16.1 Let $f = (-\Delta + |x|^2)u$. Then $f \in \mathcal{S}$ and

$$
u = \sum_{\alpha \in \mathbb{N}_0^n} \frac{1}{2|\alpha| + 1} (f, e_\alpha) e_\alpha.
$$

For all multi-indices β, we get by (15.9) a positive constant C_β such that

$$
|(f, e_\alpha)| \le C_\beta |(1+\alpha)^\beta|^{-1}.
$$

Let m be a nonnegative integer. Then for all multi-indices β, we can get by (16.1) a positive constant $C_{\beta,m}$ such that

$$
\left| \frac{1}{2|\alpha| + 1} (f, e_\alpha) \right| |\alpha|^m \le C_{\beta,m} \frac{1}{2|\alpha| + 1} |(1+\alpha)^\beta|^{-1} |\alpha|^m < \infty
$$

if we choose $|\beta|$ large enough. So, by Lemma 16.2, $u \in \mathcal{S}$.   □

The global regularity of solutions u of the Hermite equation

$$
(-\Delta + |x|^2)u = f,
$$

where f is some kind of distribution on $\mathbb{R}^n$, can best be measured by a family $\{H^{s,2} : s \in \mathbb{R}\}$ of spaces.

For every s in $(-\infty, \infty)$, we call $H^{s,2}$ the Sobolev space of order s and it is defined as the set of all distributions u in $\mathcal{S}'$ for which

$$\sum_{\alpha \in \mathbb{N}_0^n} (2|\alpha| + 1)^{2s} |u(e_\alpha)|^2 < \infty.$$

It is easy to see that for every s in $(-\infty, \infty)$, $H^{s,2}$ is a Hilbert space in which the inner product $(\,,\,)_{s,2}$ and norm $\|\,\|_{s,2}$ are given by

$$(u, v)_{s,2} = \sum_{\alpha \in \mathbb{N}_0^n} (2|\alpha| + 1)^{2s} u(e_\alpha) \overline{v(e_\alpha)}$$

and

$$\|u\|_{s,2}^2 = \sum_{\alpha \in \mathbb{N}_0^n} (2|\alpha| + 1)^{2s} |u(e_\alpha)|^2, \quad u \in H^{s,2}$$

for all u and v in $H^{s,2}$.

The following theorem is the Sobolev embedding theorem for Sobolev spaces.

Theorem 16.3 *Let s and t be real numbers such that $s \leq t$. Then*

$$H^{t,2} \subseteq H^{s,2}$$

and

$$\|u\|_{s,2} \leq \|u\|_{t,2}, \quad u \in H^{t,2}.$$

Proof Let $u \in H^{s,2}$. Then we get

$$
\begin{aligned}
\|u\|_{s,2}^2 &= \sum_{\alpha \in \mathbb{N}_0^n} (2|\alpha| + 1)^{2s} |u(e_\alpha)|^2 \\
&\leq \sum_{\alpha \in \mathbb{N}_0^n} (2|\alpha| + 1)^{2t} |u(e_\alpha)|^2 \\
&= \|u\|_{t,2}^2.
\end{aligned}
$$

$\square$

We can now give a result on the global regularity for the Hermite operator in terms of the Sobolev spaces.

Theorem 16.4 *Let $s \in (-\infty, \infty)$. Then for all f in $H^{s,2}$,*

$$(-\Delta + |x|^2)^{-1} f \in H^{s+1,2}$$

and

$$\|(-\Delta + |x|^2)^{-1} f\|_{s+1,2} \leq \|f\|_{s,2}.$$

Proof We get

$$
\begin{aligned}
\|(-\Delta + |x|^2)^{-1} f\|^2_{s+1,2} &= \sum_{\beta \in \mathbb{N}_0^n} (2|\beta| + 1)^{2s+2} \left| \sum_{\alpha \in \mathbb{N}_0^n} \frac{1}{2|\alpha| + 1} (f, e_\alpha) e_\alpha(e_\beta) \right|^2 \\
&= \sum_{\beta \in \mathbb{N}_0^n} (2|\beta| + 1)^{2s+2} \frac{1}{(2|\beta| + 1)^2} |f(e_\beta)|^2 \\
&= \|f\|^2_{s,2}.
\end{aligned}
$$

$\square$

Remark 16.5 Theorem 16.4 tells us that if $f \in H^{s,2}$, then every u in $\mathcal{S}'$ satisfying the partial differential equation

$$(-\Delta + |x|^2)u = f$$

must be in $H^{s+1,2}$, which, according to the Sobolev embedding theorem in Theorem 16.3, is one step more selective or regular than that of f. Thus, we call Theorem 16.4 a global regularity theorem. Chapter 14 of the book [58] contains results on the global regularity of pseudo-differential equations.

Historical Notes

Sergei Lvovich Sobolev (1908–1989) was a Russian mathematician. He graduated from Leningrad University (now Saint Petersburg State University) in 1929. He was instrumental in the establishment of the Novosibirsk State University in Russia. He was the first one who introduced the concept of weak solutions of partial differential equations. His most important contributions in mathematics are in Sobolev spaces, which are indispensable in the study of partial differential equations.

David Hilbert (1862–1943) was a German mathematician best known for putting forward the twenty-three open problems in mathematics at the International Congress of Mathematicians in Paris in 1900. The impact of these twenty-three problems on the development of mathematics since that time has been and is still great.

Exercises

1. Is the solution u defined by (15.11) to the partial differential equation

$$(-\Delta + |x|^2)u = f,$$

where f is a tempered distribution on $\mathbb{R}^n$, unique?

2. Prove that $H^{s,2}$ is a Hilbert space with inner product $(\ ,\ )_{s,2}$ and norm $\|\ \|_{s,2}$ given by

$$(u, v)_{s,2} = \sum_{\alpha \in \mathbb{N}_0^n} (2|\alpha| + 1)^{2s} u(e_\alpha)\overline{v(e_\alpha)}$$

and

$$\|u\|_{s,2}^2 = \sum_{\alpha \in \mathbb{N}_0^n} (2|\alpha| + 1)^{2s} |u(e_\alpha)|^2$$

for all u and v in $H^{s,2}$.

3. Let $s \in \mathbb{R}$. Prove that for all f in $H^{s,2}$,

$$\|(-\Delta + |x|^2)^{-2} f\|_{s+2} \leq \|f\|_{s,2}.$$

Chapter 17

The Heisenberg Group

The Euclidean space $\mathbb{R}^n$ has been one of the most important spaces on which mathematics is studied. It is a Riemannian manifold and it is a group with respect to the usual addition of points in $\mathbb{R}^n$. In this chapter, we introduce the Heisenberg group $\mathbb{H}^1$, which is a noncommutative group of which the underlying manifold is $\mathbb{R}^3$. In subsequent chapters, we study partial differential equations related to the Laplacians on the Heisenberg group. For the sake of simple notation, we look at the one-dimensional Heisenberg group $\mathbb{H}^1$ only. Extensions to higher-dimensional Heisenberg groups are routine. The contents in this chapter and the following chapters are based on the paper [6]. The Heisenberg group and its connections with quantum mechanics and other branches of mathematics can be found in [10, 28, 42, 48], among others.

We identify points in $\mathbb{R}^2$ with points in $\mathbb{C}$ via the identification

$$\mathbb{R}^2 \ni (x, y) \leftrightarrow z = x + iy \in \mathbb{C}.$$

Let $\mathbb{H}^1 = \mathbb{C} \times \mathbb{R}$. Then for all points (z, t) and (w, s) in $\mathbb{H}^1$, we define $(z, t) \cdot (w, s)$ by

$$(z, t) \cdot (w, s) = \left(z + w, t + s + \frac{1}{4}[z, w] \right),$$

where $[z, w]$ is the symplectic form of z and w given by

$$[z, w] = 2 \operatorname{Im}(z \overline{w}).$$

It is an exercise in this chapter to prove that $\mathbb{H}^1$ is a noncommutative group with respect to $\cdot$ in which the identity element is $(0, 0)$ and the inverse $(z, t)^{-1}$ of every element (z, t) in $\mathbb{H}^1$ is given by $(-z, -t)$. The measure $dz\, dt$ on $\mathbb{H}^1$ has the properties that for all measurable functions f on $\mathbb{H}^1$,

$$\int_{-\infty}^{\infty} \int_{\mathbb{C}} f((w, s) \cdot (z, t))\, dz\, dt = \int_{-\infty}^{\infty} \int_{\mathbb{C}} f(z, t)\, dz\, dt \qquad (17.1)$$

and

$$\int_{-\infty}^{\infty} \int_{\mathbb{C}} f((z, t) \cdot (w, s))\, dz\, dt = \int_{-\infty}^{\infty} \int_{\mathbb{C}} f(z, t)\, dz\, dt \qquad (17.2)$$

DOI: 10.1201/9781003206781-17

for all (w, s) in $\mathbb{H}^1$. Indeed, let $w = (u, v)$ and $z = (x, y)$. Then

$$\int_{-\infty}^{\infty} \int_{\mathbb{C}} f((w, s) \cdot (z, t)) \, dz \, dt$$

$$= \int_{-\infty}^{\infty} \int_{-\infty}^{\infty} \int_{-\infty}^{\infty} f\left(u + x, v + y, s + t + \frac{1}{2}(vx - uy)\right) dx \, dy \, dt.$$

Now, we make a change of variables from (x, y, t) to (ξ, η, τ) by means of the formulas

$$\xi = u + x,$$

$$\eta = v + y,$$

and

$$\tau = s + t + \frac{1}{2}(vx - uy).$$

Then

$$d\xi \, d\eta \, d\tau = J\left(\begin{array}{ccc} \xi & \eta & \tau \\ x & y & t \end{array}\right) dx \, dy \, dt,$$

where $J\left(\begin{array}{ccc} \xi & \eta & \tau \\ x & y & t \end{array}\right)$ is the Jacobian given by

$$J\left(\begin{array}{ccc} \xi & \eta & \tau \\ x & y & t \end{array}\right) = \left|\det\left(\begin{array}{ccc} \frac{\partial \xi}{\partial x} & \frac{\partial \xi}{\partial y} & \frac{\partial \xi}{\partial t} \\ \frac{\partial \eta}{\partial x} & \frac{\partial \eta}{\partial y} & \frac{\partial \eta}{\partial t} \\ \frac{\partial \tau}{\partial x} & \frac{\partial \tau}{\partial y} & \frac{\partial \tau}{\partial t} \end{array}\right)\right| = \left|\det\left(\begin{array}{ccc} 1 & 0 & 0 \\ 0 & 1 & 0 \\ v/2 & -u/2 & 1 \end{array}\right)\right| = 1$$

and hence

$$d\xi \, d\eta \, d\tau = dx \, dy \, dt.$$

Let $\zeta = (\xi, \eta)$. Then

$$\int_{-\infty}^{\infty} \int_{\mathbb{C}} f((w, s) \cdot (z, t)) \, dz \, dt = \int_{-\infty}^{\infty} \int_{\mathbb{C}} f(\zeta, \tau) \, d\zeta \, d\tau$$

and (17.1) is proved. The proof of (17.2) is the same and is best left as an exercise.

Remark 17.1 The Heisenberg group is a Lie group. The properties expressed by (17.1) and (17.2) are known, respectively, as the invariance of the Lebesgue measure with respect to the left translation and the invariance of the Lebesgue measure with respect to the right translation. It is an established fact that on any Lie group G, there is a unique measure on G that is invariant with respect to the left translation induced by the group law on G. Such a measure, which is unique up to a positive scalar multiple, is called a left Haar measure on G. A right Haar measure on G is similarly defined. A Lie group G on which the left Haar measure and the right Haar measure are equal is said to be unimodular. Thus, the Heisenberg group is unimodular.

A Lie algebra $\mathfrak{g}$ is a real vector space on which there is a binary operation $[,]$ such that $[,]$ is bilinear and the Jacobi identity holds to the effect that

$$[g_1, [g_2, g_3]] + [g_2, [g_3, g_1]] + [g_3, [g_1, g_2]] = 0$$

for all g_1, g_2, and g_3 in $\mathfrak{g}$. We call $[,]$ the bracket on $\mathfrak{g}$. To give an example of a Lie algebra, we look at vector fields on $\mathbb{H}^1$. A vector field V on $\mathbb{H}^1$ given by

$$V(x, y, t) = a(x, y, t)\frac{\partial}{\partial x} + b(x, y, t)\frac{\partial}{\partial y} + c(x, y, t)\frac{\partial}{\partial t}, \quad (x, y, t) \in \mathbb{H}^1,$$

where a, b, and c are real-valued and C^∞ functions on $\mathbb{H}^1$, is said to be left-invariant if

$$VL_{(w,s)} = L_{(w,s)}V$$

for all $(w, s) \in \mathbb{H}^1$, where $L_{(w,s)}$ is the left translation by (w, s) given by

$$(L_{(w,s)}f)(z, t) = f((w, s) \cdot (z, t)), \quad (z, t) \in \mathbb{H}^1.$$

A concrete Lie algebra is given by the following result.

Theorem 17.2 *Let $\mathfrak{h}^1$ be the set of all left-invariant vector fields on $\mathbb{H}^1$. Then $\mathfrak{h}^1$ is a Lie algebra in which the bracket $[,]$ is the commutator given by*

$$[X, Y] = XY - YX$$

for all X and Y in $\mathfrak{h}^1$.

Proof That $\mathfrak{h}^1$ is a real vector space is easy to check. Let X and Y be in $\mathfrak{h}^1$. Then we want to first show that $[X, Y] \in \mathfrak{h}^1$. To do this, write

$$X = a_1\frac{\partial}{\partial x} + b_1\frac{\partial}{\partial y} + c_1\frac{\partial}{\partial t}$$

and

$$Y = a_2\frac{\partial}{\partial x} + b_2\frac{\partial}{\partial y} + c_2\frac{\partial}{\partial t},$$

where a_1, b_1, c_1, a_2, b_2, and c_2 are real-valued and C^∞ functions on $\mathbb{H}^1$. Then a computation shows that

$$XY = a_1a_2\frac{\partial^2}{\partial x^2} + b_1b_2\frac{\partial^2}{\partial y^2} + c_1c_2\frac{\partial^2}{\partial t^2} +$$
$$(a_1b_2 + a_2b_1)\frac{\partial^2}{\partial x \partial y} + (b_1c_2 + b_2c_1)\frac{\partial^2}{\partial y \partial t} + (a_1c_2 + a_2c_1)\frac{\partial^2}{\partial t \partial x} + V_1,$$

where V_1 is a vector field on $\mathbb{H}^1$. By switching subscripts in the second-order terms in XY, we get

$$[X, Y] = XY - YX = V_1 - V_2,$$

where V_2 is another vector field on $\mathbb{H}^1$. To see that $[X,Y]$ is left-invariant, let $(w,s) \in \mathbb{H}^1$. Then we use the left invariance of X and Y to get

$$L_{(w,s)}XY = XL_{(w,s)}Y = XYL_{(w,s)}$$

and

$$L_{(w,s)}YX = YL_{(w,s)}X = YXL_{(w,s)}.$$

So,

$$[X,Y]L_{(w,s)} = L_{(w,s)}[X,Y]$$

and this proves that $[X,Y] \in \mathfrak{h}^1$. That $[\,,\,]$ is bilinear is easy to check. Finally, for all vector fields X, Y, and Z on $\mathbb{H}^1$,

$$[X,[Y,Z]] = [X,YZ - ZY] = XYZ - XZY - YZX + ZYX,$$

$$[Y,[Z,X]] = [Y,ZX - XZ] = YZX - YXZ - ZXY + XZY,$$

$$[Z,[X,Y]] = [Z,XY - YX] = ZXY - ZYX - XYZ + YXZ,$$

and therefore
$$[X,[Y,Z]] + [Y,[Z,X]] + [Z,[X,Y]] = 0,$$

i.e., the Jacobi identity holds. So, $\mathfrak{h}^1$ is a Lie algebra. $\qquad\square$

Theorem 17.3 *Let X, Y, and T be vector fields on $\mathbb{H}^1$ given by*

$$X = \frac{\partial}{\partial x} + \frac{1}{2}y\frac{\partial}{\partial t},$$

$$Y = \frac{\partial}{\partial y} - \frac{1}{2}x\frac{\partial}{\partial t},$$

and

$$T = \frac{\partial}{\partial t}.$$

Then $\{X,Y,T\}$ is a basis for $\mathfrak{h}^1$.

Proof We begin with the observation that $X \in \mathfrak{h}^1$. To see this, we have to check that
$$XL_{(w,s)} = L_{(w,s)}X$$
for all (w,s) in $\mathbb{H}^1$. Let $w = (u,v)$ and $z = (x,y)$. Then

$$(L_{(w,s)}f)(z,t) = f((w,s)\cdot(z,t)) = f\left(u+x, v+y, s+t+\frac{1}{2}(vx - uy)\right)$$

for all (z,t) in $\mathbb{H}^1$ and for all f in $C^\infty(\mathbb{H}^1)$. To simplify notation, we let

$$(\cdots) = \left(u+x, v+y, s+t+\frac{1}{2}(vx - uy)\right).$$

Thus, for all (z, t) in $\mathbb{H}^1$,

$$
\begin{aligned}
&(XL_{(w,s)}f)(z, t) \\
&= \left(\left(\frac{\partial}{\partial x} + \frac{1}{2}y\frac{\partial}{\partial t}\right)(L_{(w,s)}f)\right)(z, t) \\
&= \frac{\partial f}{\partial x}(\cdots) + \frac{1}{2}v\frac{\partial f}{\partial t}(\cdots) + \frac{1}{2}y\frac{\partial f}{\partial t}(\cdots) \\
&= \frac{\partial f}{\partial x}(\cdots) + \frac{1}{2}(v + y)\frac{\partial f}{\partial t}(\cdots)
\end{aligned}
$$

and

$$
\begin{aligned}
&(L_{(w,s)}Xf)(z, t) \\
&= (Xf)(\cdots) \\
&= \frac{\partial f}{\partial x}(\cdots) + \frac{1}{2}(v + y)\frac{\partial f}{\partial t}(\cdots).
\end{aligned}
$$

Therefore

$$
XL_{(w,s)} = L_{(w,s)}X.
$$

So, X is in $\mathfrak{h}^1$. It is an exercise to prove that Y and T are also in $\mathfrak{h}^1$. Now, a basis for the tangent space $T_{(0,0,0)}\mathbb{H}^1$ of $\mathbb{H}^1$ at the origin $(0, 0, 0)$ is given by $\left\{\frac{\partial}{\partial x}, \frac{\partial}{\partial y}, \frac{\partial}{\partial t}\right\}$. We can identify $\mathfrak{h}^1$ with $T_{(0,0,0)}\mathbb{H}^1$ as follows. To every V in $\mathfrak{h}^1$ given by

$$
V(x, y, t) = a(x, y, t)\frac{\partial}{\partial x} + b(x, y, t)\frac{\partial}{\partial y} + c(x, y, t)\frac{\partial}{\partial t},
$$

we associate the tangent vector

$$
V(0, 0, 0) = a(0, 0, 0)\frac{\partial}{\partial x} + b(0, 0, 0)\frac{\partial}{\partial y} + c(0, 0, 0)\frac{\partial}{\partial t}
$$

in $T_{(0,0,0)}\mathbb{H}^1$. Conversely, given any tangent vector

$$
\epsilon = a\frac{\partial}{\partial x} + b\frac{\partial}{\partial y} + c\frac{\partial}{\partial t}
$$

in $T_{(0,0,0)}\mathbb{H}^1$, where a, b, and c are real numbers, we let

$$
\gamma(s) = (\gamma_1(s), \gamma_2(s), \gamma_3(s)), \quad s \in \mathbb{R},
$$

be any curve in $\mathbb{H}^1$ such that

$$
\gamma(0) = (0, 0, 0)
$$

and

$$
\gamma'(0) = \epsilon.
$$

Then we associate to ϵ the vector field V^ϵ defined by

$$(V^\epsilon f)(x,y,t) = \frac{d}{ds}\bigg|_{s=0} f((x,y,t)\cdot\gamma(s)).$$

Hence for all (x,y,t) in $\mathbb{H}^1$,

$$
\begin{aligned}
&(V^\epsilon f)(x,y,t)\\
={}& \frac{d}{ds}\bigg|_{s=0} f\left(x+\gamma_1(s), y+\gamma_2(s), t+\gamma_3(s) + \frac{1}{2}(y\gamma_1(s) - x\gamma_2(s))\right)\\
={}& \gamma_1'(0)\frac{\partial f}{\partial x}(x,y,t) + \gamma_2'(0)\frac{\partial f}{\partial y}(x,y,t) +\\
& \left(\gamma_3'(0) + \frac{1}{2}y\gamma_1'(0) - \frac{1}{2}x\gamma_2'(0)\right)\frac{\partial f}{\partial t}(x,y,t)\\
={}& a\frac{\partial f}{\partial x}(x,y,t) + b\frac{\partial f}{\partial y}(x,y,t) + \left(c + \frac{1}{2}ay - \frac{1}{2}bx\right)\frac{\partial f}{\partial t}(x,y,t). \quad (17.3)
\end{aligned}
$$

So, V^ϵ depends only on ϵ. Moreover, for all (u,v,s) in $\mathbb{H}^1$, we get for all C^∞ functions f on $\mathbb{H}^1$,

$$(L_{(u,v,s)}f)(x,y,t) = f(\cdots)$$

for all (x,y,t) in $\mathbb{H}^1$. Then for all (x,y,t) in $\mathbb{H}^1$,

$$
\begin{aligned}
&(V^\epsilon L_{(u,v,s)}f)(x,y,t)\\
={}& a\frac{\partial f}{\partial x}(\cdots) + \frac{1}{2}va\frac{\partial f}{\partial t}(\cdots) +\\
& b\frac{\partial f}{\partial y}(\cdots) - \frac{1}{2}ub\frac{\partial f}{\partial t}(\cdots) +\\
& \left(c + \frac{1}{2}ya - \frac{1}{2}xb\right)\frac{\partial f}{\partial t}(\cdots)\\
={}& a\frac{\partial f}{\partial x}(\cdots) + b\frac{\partial f}{\partial y}(\cdots) +\\
& \left(c + \frac{1}{2}(v+y)a - \frac{1}{2}(u+x)b\right)\frac{\partial f}{\partial t}(\cdots).
\end{aligned}
$$

Furthermore, for all (x,y,t) in $\mathbb{H}^1$,

$$
\begin{aligned}
&(L_{(u,v,s)}V^\epsilon f)(x,y,t)\\
={}& (V^\epsilon f)(\cdots)\\
={}& a\frac{\partial f}{\partial x}(\cdots) + b\frac{\partial f}{\partial y}(\cdots) +\\
& \left(c + \frac{1}{2}(v+y)a - \frac{1}{2}(u+x)b\right)\frac{\partial f}{\partial t}(\cdots).
\end{aligned}
$$

Therefore V^ϵ is a left-invariant vector field on $\mathbb{H}^1$ and we have identified $\mathfrak{h}^1$

with the tangent space $T_{(0,0,0)}\mathbb{H}^1$ of $\mathbb{H}^1$ at the origin $(0,0,0)$. In fact, it follows from the formula (17.3) that

$$V^{\epsilon_1+\epsilon_2} = V^{\epsilon_1} + V^{\epsilon_2}, \quad \epsilon_1, \epsilon_2 \in T_{(0,0,0)}\mathbb{H}^1,$$

and

$$V^{\alpha\epsilon} = \alpha V^{\epsilon}, \quad \alpha \in \mathbb{R}, \epsilon \in T_{(0,0,0)}\mathbb{H}^1.$$

Thus, $\mathfrak{h}^1$ and $T_{(0,0,0)}\mathbb{H}^1$ are isomorphic real vector spaces and hence $\mathfrak{h}^1$ is a three-dimensional real vector space. So, we only need to prove that X, Y, and T are linearly independent. To do this, we consider the equation

$$aX + bY + cT = 0,$$

where a, b, and c are real numbers. Let f be the function on $\mathbb{H}^1$ given by

$$f(x,y,t) = x, \quad (x,y,t) \in \mathbb{H}^1.$$

Then

$$(aX + bY + cT)f = 0 \Leftrightarrow a = 0.$$

Similarly, $b = 0$ and $c = 0$. So, $\{X, Y, T\}$ is a basis for $\mathfrak{h}^1$. $\qquad\square$

The following theorem on the commutators X, Y, and T tells us that X, Y, T and their first-order commutators span the Lie algebra $\mathfrak{h}^1$. The proof is easy and left as an exercise.

Theorem 17.4 $[X,Y] = -T$ *and all the other commutators are zero.*

The choice of the vector fields X, Y, and T for a basis of $\mathfrak{h}^1$ is explained by the following theorem.

Theorem 17.5 *Let c_1, c_2, and c_3 be the coordinate axes in $\mathbb{H}^1$ and write them as*

$$c_1(s) = (s,0,0),$$
$$c_2(s) = (0,s,0),$$

and

$$c_3(s) = (0,0,s)$$

for all s in $\mathbb{R}$. Then for every C^∞ function f on $\mathbb{H}^1$,

$$(Xf)(z,t) = \left.\frac{d}{ds}\right|_{s=0} f((z,t) \cdot c_1(s)),$$

$$(Yf)(z,t) = \left.\frac{d}{ds}\right|_{s=0} f((z,t) \cdot c_2(s)),$$

and

$$(Tf)(z,t) = \left.\frac{d}{ds}\right|_{s=0} f((z,t) \cdot c_3(s))$$

for all (z,t) in $\mathbb{H}^1$.

Proof We give the proof for X only. Indeed,

$$\frac{d}{ds}\bigg|_{s=0} f((z,t) \cdot c_1(s))$$

$$= \frac{d}{ds}\bigg|_{s=0} f\left(x+s, y, t+\frac{1}{2}sy\right)$$

$$= \frac{\partial f}{\partial x}(x,y,t) + \frac{1}{2}y\frac{\partial}{\partial t}(x,y,t).$$

Thus,

$$X = \frac{\partial}{\partial x} + \frac{1}{2}y\frac{\partial}{\partial t},$$

as asserted. $\qquad\square$

Historical Notes

Werner Karl Heisenberg (1901–1976) was a German theoretical physicist best known for his works in the matrix version of quantum mechanics. The Heisenberg uncertainty principle underpins the genesis of the Heisenberg group and pseudo-differential operators in mathematics. He was the sole winner of the Nobel Prize in Physics in 1932.

Marius Sophus Lie (1842–1899) was a Norwegian mathematician. He is most remembered for his works on a branch of algebra and geometry known as Lie groups and Lie algebras, which permeate much of the modern works in mathematics and theoretical physics whenever symmetries arise.

Alfréd Haar (1885–1933) was a Hungarian mathematician best known for measures on Lie groups dubbed Haar measures. Recent research works on wavelets have revealed that the first wavelets are the Haar wavelets introduced by him in 1909.

Carl Gustav Jacob Jacobi (1801–1851), a German mathematician, made fundamental contributions to elliptic functions and number theory. The Jacobian in the change of measures due to a change of variables and the Jacobi identity in Lie algebras are attributed to him.

Exercises

1. Prove that $\mathbb{H}^1$ is a noncommutative group with respect to the multiplication $\cdot$, the identity element is $(0,0)$, and the inverse $(z,t)^{-1}$ of every element (z,t) is given by

$$(z,t)^{-1} = (-z, -t).$$

2. Prove the right invariance of the Lebesgue measure $dz\,dt$ on $\mathbb{H}^1$ as expressed by (17.2).

3. Is the space of all vector fields on $\mathbb{H}^1$ a finite-dimensional vector space?

4. Prove that the vector fields Y and T in Theorem 17.3 are left-invariant.

5. A vector field V on $\mathbb{H}^1$ is said to be right-invariant if

$$V R_{(w,s)} = R_{(w,s)} V$$

for all (w, s) in $\mathbb{H}^1$, where

$$(R_{(w,s)}f)(z,t) = f((z,t) \cdot (w,s)), \quad (z,t) \in \mathbb{H}^1,$$

for all C^∞ functions f on $\mathbb{H}^1$. Are the vector fields X, Y, and T in Theorem 17.3 right-invariant?

6. Are the vector fields $\frac{\partial}{\partial x}$ and $\frac{\partial}{\partial y}$ on $\mathbb{H}^1$ left-invariant?

7. Prove Theorem 17.4.

8. Let X_-, Y_-, and T_- be vector fields on $\mathbb{H}^1$ defined by

$$X_- = \frac{\partial}{\partial x} - \frac{1}{2}y\frac{\partial}{\partial t}, \tag{17.4}$$

$$Y_- = \frac{\partial}{\partial y} + \frac{1}{2}x\frac{\partial}{\partial t}, \tag{17.5}$$

and

$$T_- = \frac{\partial}{\partial t}.$$

Are they left-invariant? Are they right-invariant?

9. Find an explicit formula for all left-invariant vector fields on $\mathbb{H}^1$. Do the same for all right-invariant vector fields on $\mathbb{H}^1$.

10. A vector field V on $\mathbb{H}^1$ is said to be bi-invariant if it is both left-invariant and right-invariant. Find an explicit formula for all bi-invariant vector fields on $\mathbb{H}^1$.

.

Chapter 18

The Sub-Laplacian and the Twisted Laplacians

By Theorem 17.4 in the preceding chapter, the vector fields X and Y on the Heisenberg group $\mathbb{H}^1$ enjoy the special property that

$$[X, Y] = -T.$$

As such, by Theorem 17.3, the vector fields X and Y and their first-order commutators span the Lie algebra $\mathfrak{h}^1$. So, the vector fields X and Y play a special role in the study of the Heisenberg group. They are known as the horizontal vector fields on $\mathbb{H}^1$ and T is referred to as the missing direction.

The sub-Laplacian $\mathcal{L}$ on $\mathbb{H}^1$ is defined by

$$\mathcal{L} = -(X^2 + Y^2). \tag{18.1}$$

To see the sub-Laplacian $\mathcal{L}$ more explicitly, we note that

$$
\begin{aligned}
X^2 &= \left(\frac{\partial}{\partial x} + \frac{1}{2} y \frac{\partial}{\partial t} \right) \left(\frac{\partial}{\partial x} + \frac{1}{2} y \frac{\partial}{\partial t} \right) \\
&= \frac{\partial^2}{\partial x^2} + y \frac{\partial^2}{\partial x \partial t} + \frac{1}{4} y^2 \frac{\partial^2}{\partial t^2}
\end{aligned}
$$

and

$$
\begin{aligned}
Y^2 &= \left(\frac{\partial}{\partial y} - \frac{1}{2} x \frac{\partial}{\partial t} \right) \left(\frac{\partial}{\partial y} - \frac{1}{2} x \frac{\partial}{\partial t} \right) \\
&= \frac{\partial^2}{\partial y^2} - x \frac{\partial^2}{\partial y \partial t} + \frac{1}{4} x^2 \frac{\partial^2}{\partial t^2}.
\end{aligned}
$$

Therefore

$$\mathcal{L} = -\Delta - \frac{1}{4}(x^2 + y^2)\frac{\partial^2}{\partial t^2} + \left(x \frac{\partial}{\partial y} - y \frac{\partial}{\partial x} \right) \frac{\partial}{\partial t},$$

where

$$\Delta = \frac{\partial^2}{\partial x^2} + \frac{\partial^2}{\partial y^2}.$$

Let $\frac{\partial}{\partial z}$ and $\frac{\partial}{\partial \bar{z}}$ be linear partial differential operators on $\mathbb{R}^2$ given by

$$\frac{\partial}{\partial z} = \frac{\partial}{\partial x} - i \frac{\partial}{\partial y}$$

DOI: 10.1201/9781003206781-18

and

$$\frac{\partial}{\partial \overline{z}} = \frac{\partial}{\partial x} + i\frac{\partial}{\partial y}.$$

Let $\tau \in \mathbb{R} \setminus \{0\}$. Then we define the partial differential operators Z_τ and $\overline{Z}_\tau$ by

$$Z_\tau = \frac{\partial}{\partial z} + \frac{1}{2}\tau\overline{z}, \quad \overline{z} = x - iy,$$

and

$$\overline{Z}_\tau = \frac{\partial}{\partial \overline{z}} - \frac{1}{2}\tau z, \quad z = x + iy.$$

In fact, $-\overline{Z}_\tau$ is the formal adjoint of Z_τ, i.e.,

$$(Z_\tau \varphi, \psi) = -(\varphi, \overline{Z}_\tau \psi)$$

for all Schwartz functions φ and ψ on $\mathbb{R}^2$.

Let L_τ be the linear partial differential operator on $\mathbb{R}^2$ defined by

$$L_\tau = -\frac{1}{2}(Z_\tau\overline{Z}_\tau + \overline{Z}_\tau Z_\tau).$$

In order to write down an explicit formula for L_τ, we note that

$$
\begin{aligned}
Z_\tau\overline{Z}_\tau &= \left(\frac{\partial}{\partial z} + \frac{1}{2}\tau\overline{z}\right)\left(\frac{\partial}{\partial \overline{z}} - \frac{1}{2}\tau z\right) \\
&= \Delta - \frac{1}{2}\tau\left(z\frac{\partial}{\partial z} + 1\right) + \frac{1}{2}\tau\overline{z}\frac{\partial}{\partial \overline{z}} - \frac{1}{4}\tau^2|z|^2
\end{aligned}
$$

and

$$
\begin{aligned}
\overline{Z}_\tau Z_\tau &= \left(\frac{\partial}{\partial \overline{z}} - \frac{1}{2}\tau z\right)\left(\frac{\partial}{\partial z} + \frac{1}{2}\tau\overline{z}\right) \\
&= \Delta + \frac{1}{2}\tau\left(\overline{z}\frac{\partial}{\partial \overline{z}} + 1\right) - \frac{1}{2}\tau z\frac{\partial}{\partial z} - \frac{1}{4}\tau^2|z|^2.
\end{aligned}
$$

So,

$$
\begin{aligned}
L_\tau &= -\frac{1}{2}(Z_\tau\overline{Z}_\tau + \overline{Z}_\tau Z_\tau) \\
&= -\Delta + \frac{1}{2}\tau z\frac{\partial}{\partial z} - \frac{1}{2}\tau\overline{z}\frac{\partial}{\partial \overline{z}} + \frac{1}{4}|z|^2\tau^2.
\end{aligned}
$$

Since

$$z\frac{\partial}{\partial z} = x\frac{\partial}{\partial x} + y\frac{\partial}{\partial y} - i\left(x\frac{\partial}{\partial y} - y\frac{\partial}{\partial x}\right)$$

and

$$\overline{z}\frac{\partial}{\partial \overline{z}} = x\frac{\partial}{\partial x} + y\frac{\partial}{\partial y} + i\left(x\frac{\partial}{\partial y} - y\frac{\partial}{\partial x}\right),$$

it follows that

$$L_\tau = -\Delta + \frac{1}{4}(x^2 + y^2)\tau^2 - i\left(x\frac{\partial}{\partial y} - y\frac{\partial}{\partial x}\right)\tau. \qquad (18.2)$$

Thus, L_τ is the Hermite operator $-\Delta + \frac{1}{4}(x^2 + y^2)\tau^2$ perturbed by the partial differential operator $-iN\tau$, where

$$N = x\frac{\partial}{\partial y} - y\frac{\partial}{\partial x}$$

is the rotation operator. We call L_τ the twisted Laplacian with parameter τ. We denote L_1 by L, and the corresponding constituent operators Z_1 and $\overline{Z}_1$ in L_1 by Z and $\overline{Z}$, respectively.

To see the connection between the sub-Laplacian and the twisted Laplacians, we define for every Schwartz function f on $\mathbb{H}^1$, the function f^τ, $\tau \in \mathbb{R} \setminus \{0\}$, on $\mathbb{C}$ by

$$f^\tau(z) = (2\pi)^{-1/2} \int_{-\infty}^{\infty} e^{it\tau} f(z,t)\, dt, \quad z \in \mathbb{C}.$$

It should be noted that $f^\tau(z)$ is in fact the inverse Fourier transform of $f(z,t)$ with respect to t evaluated at τ. Then we have the following connection.

Theorem 18.1 *Let φ be a Schwartz function on $\mathbb{H}^1$. Then for all $\tau \in \mathbb{R} \setminus \{0\}$,*

$$(\mathcal{L}\varphi)^\tau(z) = (L_\tau \varphi^\tau)(z), \quad z \in \mathbb{C}.$$

Proof We begin with the simple observation that for all Schwartz functions f on $\mathbb{R}$ and all $\tau \in \mathbb{R} \setminus \{0\}$,

$$(2\pi)^{-1/2} \int_{-\infty}^{\infty} e^{it\tau} f'(t)\, dt$$

$$= (2\pi)^{-1/2} \left(f(t)e^{it\tau}\big|_0^\infty - \int_{-\infty}^{\infty} i\tau e^{it\tau} f(t)\, dt \right).$$

Thus,

$$(2\pi)^{-1/2} \int_{-\infty}^{\infty} e^{it\tau} f'(t)\, dt = -i\tau \check{f}(\tau).$$

So, replacing $\frac{\partial}{\partial t}$ in $\mathcal{L}$ by $-i\tau$, the result follows. $\qquad \square$

We end this chapter with a discussion on the ellipticity of the most general linear partial differential operator $P(x,D)$ with C^∞ coefficients on $\mathbb{R}^n$ and order m. In fact,

$$P(x,D) = \sum_{|\alpha| \le m} a_\alpha(x) D^\alpha,$$

where $a_\alpha \in C^\infty(\mathbb{R}^n)$. Replacing D by ξ, we get the symbol $P(x, \xi)$ of the operator $P(x, D)$ given by

$$P(x, \xi) = \sum_{|\alpha| \leq m} a_\alpha(x)\xi^\alpha, \quad \xi \in \mathbb{R}^n.$$

The principal symbol $P_m(x, \xi)$ of $P(x, D)$ is defined by

$$P_m(x, \xi) = \sum_{|\alpha| = m} a_\alpha(x)\xi^\alpha, \quad \xi \in \mathbb{R}^n.$$

Let $x_0 \in \mathbb{R}^n$. Then $P(x, D)$ is said to be elliptic at x_0 if

$$P_m(x_0, \xi) = 0, \, \xi \in \mathbb{R}^n \Rightarrow \xi = 0. \tag{18.3}$$

The operator $P(x, D)$ is said to be elliptic on $\mathbb{R}^n$ if it is elliptic at every point in $\mathbb{R}^n$. One of the best known properties of ellipticity on $\mathbb{R}^n$ is that

$$u \in \mathcal{D}'(\mathbb{R}^n), \, P(x, D)u \in C^\infty(\mathbb{R}^n) \Rightarrow u \in C^\infty(\mathbb{R}^n), \tag{18.4}$$

where $\mathcal{D}'(\mathbb{R}^n)$ is the space of all distributions on $\mathbb{R}^n$.

Remark 18.2 A precise description of $\mathcal{D}'(\mathbb{R}^n)$ is desirable. In fact, we can give a description of the more general space $\mathcal{D}'(\Omega)$, where Ω is an open and connected subset of $\mathbb{R}^n$. Let $C_0^\infty(\Omega)$ be the set of all C^∞ functions φ on Ω such that

$$\text{supp}(\varphi) \subset \Omega.$$

Let $\{\varphi_j\}_{j=1}^\infty$ be a sequence in $C_0^\infty(\Omega)$ such that there exists a compact subset K of Ω for which

$$\text{supp}(\varphi_j) \subseteq K, \quad j = 1, 2, \ldots,$$

and for all multi-indices α,

$$D^\alpha \varphi_j \to 0$$

uniformly on Ω as $j \to \infty$. Then we say that $\varphi_j \to 0$ in $C_0^\infty(\Omega)$ as $j \to \infty$. A linear functional T on $C_0^\infty(\Omega)$ is said to be a distribution on Ω if

$$T(\varphi_j) \to 0$$

for all sequences $\{\varphi_j\}_{j=1}^\infty$ such that

$$\varphi_j \to 0$$

in $C_0^\infty(\Omega)$ as $j \to \infty$. The set of all distributions on Ω is denoted by $\mathcal{D}'(\Omega)$. It should be noted that $\mathcal{D}'(\mathbb{R}^n)$ is larger than the space of all tempered distributions on $\mathbb{R}^n$.

It should be noted that ellipticity on $\mathbb{R}^n$ implies (18.4), but the converse is false. In fact, the operator $P(x, D)$ is said to be hypoelliptic on $\mathbb{R}^n$ if (18.4) is valid. This notion of hypoellipticity and the notion of global hypoellipticty in Chapters 10 and 16 are different notions in the global regularity of partial differential equations. A systematic study on the hypoellipticity of general linear partial differential operators can be traced back to the work [22] of Hörmander. The books [35, 36, 38] and many others are excellent accounts on hypoellipticity.

Theorem 18.3 *Let $\tau \in \mathbb{R} \setminus \{0\}$. Then L_τ is elliptic on $\mathbb{R}^2$.*

Proof By (18.2), the principal symbol of L_τ is the function $|\xi|^2$ on $\mathbb{R}^2$. So, (18.3) is valid at every point in $\mathbb{R}^2$. Therefore L_τ is elliptic on $\mathbb{R}^2$. $\qquad \square$

Theorem 18.4 *The sub-Laplacian $\mathcal{L}$ is nowhere elliptic on $\mathbb{R}^3$.*

Proof Using the formula (18.1) and the replacement of $\frac{\partial}{\partial x}$, $\frac{\partial}{\partial y}$, and $\frac{\partial}{\partial t}$ by $i\xi$, $i\eta$, and $i\lambda$, respectively, the symbol $P(x, y, t, \xi, \eta, \lambda)$ of $\mathcal{L}$ is given by

$$
\begin{aligned}
P(x, y, t, \xi, \eta, \lambda) &= -\left(i\xi + \frac{1}{2}yi\lambda\right)^2 - \left(i\eta - \frac{1}{2}xi\lambda\right)^2 \\
&= \left(\xi + \frac{1}{2}y\lambda\right)^2 + \left(\eta - \frac{1}{2}x\lambda\right)^2.
\end{aligned}
$$

Let $(x_0, y_0, t_0) \in \mathbb{R}^3$. If $x_0 = y_0 = 0$, then $P(x_0, y_0, t_0, \xi, \eta, \lambda) = 0$ for all points $(0, 0, \lambda)$, $\lambda \in \mathbb{R}$. Suppose that $x_0 \neq 0$ or $y_0 \neq 0$. To be specific, let us assume that $x_0 \neq 0$. Then $P(x_0, y_0, t_0, \xi, \eta, \lambda) = 0$ for all (ξ, η, λ) such that

$$
\begin{cases}
\xi = -\frac{1}{2}y_0\lambda, \\
\eta = \frac{1}{2}x_0\lambda.
\end{cases}
$$

This completes the proof that $\mathcal{L}$ is nowhere elliptic on $\mathbb{R}^3$. $\qquad \square$

Remark 18.5 Applying a result of Hörmander in [21] in our context of the Heisenberg group $\mathbb{H}^1$ to the effect that if we have a sum of squares of vector fields such that the vector fields and their commutators up to a certain order span the Lie algebra $\mathfrak{h}^1$, then the sum of squares is hypoelliptic on $\mathbb{R}^3$. So, by Hörmander's result and Theorem 17.4, the sub-Laplacian $\mathcal{L}$ is hypoelliptic.

Historical Notes

Lars Hörmander (1931–2012) was a Swedish mathematician. His works on the theory of general partial differential operators started with his Lund University doctoral dissertation [20] published in *Acta Mathematica* in 1955 and culminated in the magnificent four-volume treatise [23]–[26] titled *The Analysis of Linear Partial Differential Operators* published in the 1980s. He won a Fields Medal for his works in the modern theory of partial differential equations in 1962.

Exercises

1. Prove that the sub-Laplacian $\mathcal{L}$ is left-invariant on the Heisenberg group $\mathbb{H}^1$, *i.e.*,

$$\mathcal{L}L_{(w,s)} = L_{(w,s)}\mathcal{L}, \quad (w,s) \in \mathbb{H}^1.$$

2. Let $\mathcal{L}_- = -(X_-^2 + Y_-^2)$, where X_- and Y_- are given, respectively, by (17.4) and (17.5). Give the explicit formula for $\mathcal{L}_-$. How is it related to the sub-Laplacian $\mathcal{L}$?

3. Prove that $\mathcal{L}_-$ is right-invariant on the Heisenberg group $\mathbb{H}^1$.

4. What is the analog of the formula (18.2) for the twisted Laplacians of $\mathcal{L}_-$?

5. Let $\tau \in \mathbb{R} \setminus \{0\}$. Prove that

$$(Z_\tau\varphi, \psi) = -(\varphi, \overline{Z}_\tau\psi)$$

for all Schwartz functions φ and ψ on $\mathbb{R}^2$.

6. Let $\tau \in \mathbb{R} \setminus \{0\}$. Determine whether or not the operator $L_\tau L_{-\tau}$ is elliptic on $\mathbb{R}^2$.

7. Is the operator $\mathcal{L}_-$ hypoelliptic on $\mathbb{R}^3$?

8. Is the operator $\mathcal{L}\mathcal{L}_-$ hypoelliptic on $\mathbb{R}^3$?

Chapter 19

Convolutions on the Heisenberg Group

We are interested in solving for u of the initial value problem for the heat equation governed by the sub-Laplacian $\mathcal{L}$, i.e.,

$$\begin{cases} \frac{\partial u}{\partial \rho}(z,t,\rho) = -(\mathcal{L}u)(z,t,\rho), \\ u(z,t,0) = f(z,t) \end{cases}$$

for all $(z,t) \in \mathbb{H}^1$ and all time $\rho > 0$. Formally, u is given by

$$u(z,t,\rho) = (e^{-\rho\mathcal{L}}f)(z,t), \quad (z,t) \in \mathbb{H}^1, \, \rho > 0.$$

It should be observed that from now on we use ρ to denote time because the usual letter t for time has been used to denote the missing direction in the Heisenberg group.

We are also geared to find a solution u of the Poisson equation

$$\mathcal{L}u = f,$$

where f is a given suitable function on $\mathbb{H}^1$. Formally, u is given by

$$u = \mathcal{L}^{-1}f.$$

In order to determine $e^{-\rho\mathcal{L}}f$ and $\mathcal{L}^{-1}f$ explicitly, we first introduce the notion of a convolution on the Heisenberg group $\mathbb{H}^1$. Let f and g be measurable functions on $\mathbb{H}^1$. Then we define the convolution $f *_{\mathbb{H}^1} g$ of f and g on $\mathbb{H}^1$ by

$$(f *_{\mathbb{H}^1} g)(z,t) = \int_{-\infty}^{\infty} \int_{\mathbb{C}} f((z,t) \cdot (w,s)^{-1}) g(w,s) \, dw \, ds$$

provided that the integral exists.

Our goal is to find a formula for the heat kernel K_ρ, $\rho > 0$, of $\mathcal{L}$ for which

$$e^{-\rho\mathcal{L}}f = f *_{\mathbb{H}^1} K_\rho \tag{19.1}$$

for suitable functions f on $\mathbb{H}^1$. An equally important goal is to find a formula for the Green function $\mathcal{G}$ of $\mathcal{L}$ such that

$$\mathcal{L}^{-1}f = f *_{\mathbb{H}^1} \mathcal{G}$$

DOI: 10.1201/9781003206781-19

for all suitable functions f on $\mathbb{H}^1$.

Let $\lambda \in \mathbb{R}$. Then we define the twisted convolution $f *_\lambda g$ of two measurable functions f and g on $\mathbb{C}$ by

$$(f *_\lambda g)(z) = \int_\mathbb{C} f(z-w)g(w)e^{i\lambda[z,w]}dw, \quad z \in \mathbb{C}, \tag{19.2}$$

provided that the integral exists. The relationship between the convolution on $\mathbb{H}^1$ and the twisted convolution is given by the following theorem.

Theorem 19.1 *Let f and g be functions in $L^1(\mathbb{H}^1)$. Then*

$$(f *_{\mathbb{H}^1} g)^\tau = (2\pi)^{1/2}(f^\tau *_{\tau/4} g^\tau).$$

Proof For all z in $\mathbb{C}$,

$$
\begin{aligned}
&(f *_{\mathbb{H}^1} g)^\tau(z) \\
={}& (2\pi)^{-1/2} \int_{-\infty}^\infty e^{it\tau}(f *_{\mathbb{H}^1} g)(z,t)\,dt \\
={}& (2\pi)^{-1/2} \int_{-\infty}^\infty e^{it\tau} \left(\int_{-\infty}^\infty \int_\mathbb{C} f\left(z-w, t-s-\frac{1}{4}[z,w]\right) g(w,s)\,dw\,ds \right) dt.
\end{aligned}
$$

Let $t' = t - \frac{1}{4}[z,w]$. Then

$$
\begin{aligned}
&(f *_{\mathbb{H}^1} g)^\tau(z) \\
={}& (2\pi)^{-1/2} \int_{-\infty}^\infty \int_{-\infty}^\infty \int_\mathbb{C} e^{it'\tau} f(z-w, t'-s)g(w,s)e^{i\frac{\tau}{4}[z,w]}dw\,ds\,dt'.
\end{aligned}
$$

On the other hand, for all z in $\mathbb{C}$, we get

$$
\begin{aligned}
&(f^\tau *_{\tau/4} g^\tau)(z) \\
={}& \int_\mathbb{C} f^\tau(z-w)g^\tau(w)e^{i\frac{\tau}{4}[z,w]}dw \\
={}& (2\pi)^{-1/2} \int_\mathbb{C} \left(\int_{-\infty}^\infty f(z-w, \cdot - s)g(w,s)ds \right)^\vee (\tau)e^{i\frac{\tau}{4}[z,w]}dw \\
={}& (2\pi)^{-1} \int_\mathbb{C} \int_{-\infty}^\infty e^{it\tau} \left(\int_{-\infty}^\infty f(z-w, t-s)g(w,s)\,ds \right) e^{i\frac{\tau}{4}[z,w]}dt\,dw,
\end{aligned}
$$

and the proof is complete by Fubini's theorem. $\qquad\square$

Using Theorem 18.1, Theorem 19.1, and the definition of the heat kernel given by (19.1), we see that for positive time ρ and suitable functions f on $\mathbb{H}^1$,

$$e^{-\rho L_\tau}f^\tau = (2\pi)^{1/2}(f^\tau *_{\tau/4} K_\rho^\tau) = (2\pi)^{1/2}(K_\rho^\tau *_{-\tau/4} f^\tau), \quad \tau \in \mathbb{R}\setminus\{0\}. \tag{19.3}$$

So, if we can compute the heat kernel $e^{-i\frac{\tau}{4}[z,w]}K_\rho^\tau(z,w)$, $z,w \in \mathbb{C}$, $\rho > 0$, of the twisted Laplacian L_τ, then the heat kernel K_ρ of $\mathcal{L}$ can be obtained by taking the Fourier transform of K_ρ^τ with respect to τ.

Similarly,

$$L_\tau^{-1}f^\tau = (2\pi)^{1/2}(f^\tau *_{\tau/4} \mathcal{G}^\tau) = (2\pi)^{1/2}(\mathcal{G}^\tau *_{-\tau/4} f^\tau), \quad \tau \in \mathbb{R} \setminus \{0\}. \quad (19.4)$$

So, the Green function $\mathcal{G}$ of $\mathcal{L}$ can be obtained by taking the Fourier transform of $\mathcal{G}^\tau$ with respect to τ once $\mathcal{G}^\tau$, $\tau \in \mathbb{R} \setminus \{0\}$, is computed.

In order to find the heat kernel and the Green function of L_τ for τ in $\mathbb{R} \setminus \{0\}$, we need Wigner transforms and Weyl transforms, which we study in the following chapter.

Exercises

1. Derive the formula for $e^{-\rho\mathcal{L}}f$ explicitly in terms of an integral, K_ρ, and f.

2. Derive the formula for $\mathcal{L}^{-1}f$ explicitly in terms of an integral, $\mathcal{G}$, and f.

3. Prove formula (19.3).

4. Prove formula (19.4).

Chapter 20

Wigner Transforms and Weyl Transforms

Wigner transforms and Weyl transforms are not only useful as tools in this book, but also arise in many different contexts in mathematics. For this reason, we find it worthwhile to present them in the setting of $\mathbb{R}^n$. The books [10, 54] contain much more information on these topics.

We identify points in $\mathbb{R}^n \times \mathbb{R}^n$ with points in $\mathbb{C}^n$ via the identification

$$\mathbb{R}^n \times \mathbb{R}^n \ni (q, p) \leftrightarrow q + ip \in \mathbb{C}^n.$$

Let q and p be points in $\mathbb{R}^n$, and let f be a measurable function on $\mathbb{R}^n$. We define the function $\rho(q, p)f$ on $\mathbb{R}^n$ by

$$(\rho(q, p)f)(x) = e^{iq \cdot x + \frac{1}{2}iq \cdot p} f(x + p), \quad x \in \mathbb{R}^n.$$

Let f and g be functions in $\mathcal{S}$. Then we define the Fourier–Wigner transform $V(f, g)$ of f and g on $\mathbb{R}^n$ to be the function on $\mathbb{R}^n \times \mathbb{R}^n$ by

$$V(f, g)(q, p) = (2\pi)^{-n/2}(\rho(q, p)f, g), \quad q, p \in \mathbb{R}^n.$$

The following theorem, which can be found in [54], gives another formula for the Fourier–Wigner transform.

Theorem 20.1 *The Fourier–Wigner transform $V(f, g)$ of Schwartz functions f and g on $\mathbb{R}^n$ is given by*

$$V(f, g)(q, p) = (2\pi)^{-n/2} \int_{\mathbb{R}^n} e^{iq \cdot y} f\left(y + \frac{p}{2}\right) \overline{g\left(y - \frac{p}{2}\right)} \, dy$$

for all q and p in $\mathbb{R}^n$.

Proof For all q and p in $\mathbb{R}^n$, we have

$$V(f, g)(q, p) = (2\pi)^{-n/2} \int_{\mathbb{R}^n} e^{iq \cdot x + \frac{1}{2}iq \cdot p} f(x + p) \overline{g(x)} \, dx.$$

Let us change the variable x to y via

$$x = y - \frac{p}{2}.$$

DOI: 10.1201/9781003206781-20

Then

$$V(f,g)(q,p) = (2\pi)^{-n/2} \int_{\mathbb{R}^n} e^{iq\cdot y} f\left(y + \frac{p}{2}\right) \overline{g\left(y - \frac{p}{2}\right)} dy.$$

$\square$

Let f and g be functions in $\mathcal{S}$. Then the Wigner transform $W(f,g)$ of f and g is defined by

$$W(f,g) = V(f,g)^{\wedge}.$$

Theorem 20.2 *For all Schwartz functions f and g on $\mathbb{R}^n$,*

$$W(f,g)(x,\xi) = (2\pi)^{-n/2} \int_{\mathbb{R}^n} e^{-i\xi\cdot p} f\left(x + \frac{p}{2}\right) \overline{g\left(x - \frac{p}{2}\right)} dp, \quad x,\xi \in \mathbb{R}^n.$$

Proof By the definition of the Wigner transform, we have for all x and ξ in $\mathbb{R}^n$,

$$
\begin{aligned}
& W(f,g)(x,\xi) \\
= \; & (2\pi)^{-3n/2} \int_{\mathbb{R}^n} \int_{\mathbb{R}^n} e^{-ix\cdot q - i\xi\cdot p} \left(\int_{\mathbb{R}^n} e^{iq\cdot y} f\left(y + \frac{p}{2}\right) \overline{g\left(y - \frac{p}{2}\right)} dy\right) dq\, dp \\
= \; & (2\pi)^{-3n/2} \int_{\mathbb{R}^n} \int_{\mathbb{R}^n} \left(\int_{\mathbb{R}^n} e^{-iq\cdot(x-y)} dq\right) e^{-i\xi\cdot p} f\left(y + \frac{p}{2}\right) \overline{g\left(y - \frac{p}{2}\right)} dy\, dp \\
= \; & (2\pi)^{-n/2} \int_{\mathbb{R}^n} e^{-i\xi\cdot p} \left(\int_{\mathbb{R}^n} \delta(x - y) f\left(y + \frac{p}{2}\right) \overline{g\left(y - \frac{p}{2}\right)} dy\right) dp \\
= \; & (2\pi)^{-n/2} \int_{\mathbb{R}^n} e^{-i\xi\cdot p} f\left(x + \frac{p}{2}\right) \overline{g\left(x - \frac{p}{2}\right)} dp.
\end{aligned}
$$

$\square$

In the above equations, we have used the facts that

$$(2\pi)^{-n/2} \int_{\mathbb{R}^n} e^{iq\cdot(y-x)} dq = \delta(y - x) = \delta(x - y)$$

and

$$\int_{\mathbb{R}^n} \delta(x - y) f(y)\, dy = f(x).$$

A simple fact about the Wigner transform is given by the following proposition, and its proof is left as an exercise.

Proposition 20.3 *Let f and g be functions in $L^2(\mathbb{R}^n)$. Then*

$$W(g,f) = \overline{W(f,g)}.$$

The following Moyal identity can be thought of as the Plancherel formula for Wigner transforms.

Theorem 20.4 *For all functions f_1, f_2, g_1, and g_2 in $L^2(\mathbb{R}^n)$,*

$$(W(f_1, g_1), W(f_2, g_2)) = (f_1, f_2)\overline{(g_1, g_2)}.$$

Proof By Plancherel's theorem, we have

$$(W(f_1, g_1), W(f_2, g_2))$$
$$= \int_{\mathbb{R}^n} \int_{\mathbb{R}^n} f_1\left(x + \frac{p}{2}\right)\overline{g_1\left(x - \frac{p}{2}\right)}\overline{f_2\left(x + \frac{p}{2}\right)}g_2\left(x - \frac{p}{2}\right) dx\, dp.$$

Let $u = x + \frac{p}{2}$ and $v = x - \frac{p}{2}$. Then

$$du\, dv = \left|\det\begin{pmatrix} I & \frac{1}{2}I \\ I & -\frac{1}{2}I \end{pmatrix}\right| dx\, dp = dx\, dp,$$

where I is the identity matrix of order $n \times n$. Hence

$$(W(f_1, g_1), W(f_2, g_2))$$
$$= \int_{\mathbb{R}^n} \int_{\mathbb{R}^n} f_1(u)\overline{g_1(v)}\,\overline{f_2(u)}g_2(v)\, du\, dv$$
$$= (f_1, f_2)\overline{(g_1, g_2)},$$

as asserted. $\qquad\square$

We are now in the position to define Weyl transforms. Let σ be a function in $L^2(\mathbb{R} \times \mathbb{R}^n)$. Then for all functions f in $L^2(\mathbb{R}^n)$, the Weyl transform $W_\sigma f$ of f with symbol σ is given by

$$(W_\sigma f, g) = (2\pi)^{-n/2} \int_{\mathbb{R}^n} \int_{\mathbb{R}^n} \sigma(x, \xi) W(f, g)(x, \xi)\, dx\, d\xi, \quad g \in L^2(\mathbb{R}^n).$$

In fact, by the adjoint formula for the Fourier transform, we have for all f in $L^2(\mathbb{R}^n)$,

$$\begin{aligned}
(W_\sigma f, g) &= (2\pi)^{-n/2} \int_{\mathbb{R}^n} \int_{\mathbb{R}^n} \sigma(q, p) V(f, g)^\wedge(q, p)\, dq\, dp \\
&= (2\pi)^{-n/2} \int_{\mathbb{R}^n} \int_{\mathbb{R}^n} \hat{\sigma}(q, p) V(f, g)(q, p)\, dq\, dp \\
&= (2\pi)^{-n} \int_{\mathbb{R}^n} \int_{\mathbb{R}^n} \hat{\sigma}(q, p)(\rho(q, p)f, g)\, dq\, dp
\end{aligned}$$

and hence

$$W_\sigma f = (2\pi)^{-n} \int_{\mathbb{R}^n} \int_{\mathbb{R}^n} \hat{\sigma}(q, p)\rho(q, p)f\, dq\, dp. \tag{20.1}$$

We end this chapter with a product formula in [17] for the Weyl transforms. We need a lemma.

Lemma 20.5 *let z and w be points in $\mathbb{C}^n$. Then*

$$\rho(z)\rho(w) = \rho(z+w)e^{\frac{1}{4}i[z,w]}.$$

Here, $[z,w]$ is the symplectic form of z and w defined by

$$[z,w] = 2\operatorname{Im}(z\cdot\overline{w}), \quad z,w \in \mathbb{C}^n,$$

where

$$z = (z_1, z_2, \ldots, z_n),$$
$$w = (w_1, w_2, \ldots, w_n),$$

and

$$z\cdot\overline{w} = \sum_{j=1}^{n} z_j \overline{w_j}.$$

Proof Let $z = q + ip$ and $w = v + iu$. Then we get for all f in $L^2(\mathbb{R}^n)$,

$$
\begin{aligned}
&(\rho(z)\rho(w)f)(x)\\
=\ & e^{iq\cdot x + \frac{1}{2}iq\cdot p}(\rho(w)f)(x+p)\\
=\ & e^{iq\cdot x + \frac{1}{2}iq\cdot p}e^{iv\cdot(x+p) + \frac{1}{2}iv\cdot u}f(x+p+u)\\
=\ & e^{iq\cdot x + \frac{1}{2}iq\cdot p + iv\cdot x + iv\cdot p + \frac{1}{2}iv\cdot u}f(x+p+u)
\end{aligned}
$$

for all x in $\mathbb{R}^n$. On the other hand,

$$
\begin{aligned}
&(\rho(z+w)e^{\frac{1}{4}i[z,w]}f)(x)\\
=\ & e^{\frac{1}{2}ip\cdot v - \frac{1}{2}iq\cdot u + i(q+v)\cdot x + \frac{1}{2}i(q+v)\cdot(p+u)}f(x+p+u)\\
=\ & e^{\frac{1}{2}ip\cdot v - \frac{1}{2}iq\cdot u + iq\cdot x + iv\cdot x + \frac{1}{2}iq\cdot p + \frac{1}{2}iq\cdot u + \frac{1}{2}iv\cdot p + \frac{1}{2}iv\cdot u}f(x+p+u)\\
=\ & e^{ip\cdot v + iq\cdot x + iv\cdot x + \frac{1}{2}iq\cdot p + \frac{1}{2}iv\cdot u}f(x+p+u)
\end{aligned}
$$

for all x in $\mathbb{R}^n$, and this completes the proof. $\qquad\square$

The twisted convolution of two measurable functions on $\mathbb{C}^n$ can be defined as in the case of (19.2). Indeed, let f and g be measurable functions on $\mathbb{C}^n$. Then for $\lambda \in \mathbb{R} \setminus \{0\}$, the twisted convolution $f *_\lambda g$ of f and g is defined by

$$(f *_\lambda g)(z) = \int_{\mathbb{C}^n} f(z-w)\,g(w)e^{i\lambda[z,w]}dw, \quad z \in \mathbb{C}^n,$$

provided that the integral exists.

Theorem 20.6 *Let σ and τ be functions in $L^2(\mathbb{R}^n \times \mathbb{R}^n)$. Then we have*

$$W_\sigma W_\tau = W_\omega,$$

where

$$\hat{\omega} = (2\pi)^{-n}(\hat{\sigma} *_{1/4} \hat{\tau}).$$

Proof Let $z = (q, p) \in \mathbb{C}^n$. Then by (20.1), we have for all f in $L^2(\mathbb{R}^n)$,

$$(W_\sigma(W_\tau f))(x) = (2\pi)^{-n} \int_{\mathbb{C}^n} \hat{\sigma}(z)(\rho(z)W_\tau f)(x)\,dz, \quad x \in \mathbb{R}^n.$$

By Lemma 20.5, we get

$$(\rho(z)W_\tau f)(x)$$
$$= e^{iq\cdot x + \frac{1}{2}iq\cdot p}(2\pi)^{-n} \int_{\mathbb{C}^n} \hat{\tau}(w)(\rho(w)f)(x + p)\,dw$$
$$= (2\pi)^{-n} \int_{\mathbb{C}^n} \hat{\tau}(w)(\rho(z)\rho(w)f)(x)\,dw$$
$$= (2\pi)^{-n} \int_{\mathbb{C}^n} \hat{\tau}(w)(\rho(z + w)e^{\frac{1}{4}i[z,w]}f)(x)\,dw$$

for all x in $\mathbb{R}^n$. Thus, for all x in $\mathbb{R}^n$,

$$(W_\sigma(W_\tau f))(x)$$
$$= (2\pi)^{-2n} \int_{\mathbb{C}^n} \hat{\sigma}(z)\left(\int_{\mathbb{C}^n} \hat{\tau}(w)(\rho(z + w)e^{\frac{1}{4}i[z,w]}f)(x)\,dw\right) dz$$
$$= (2\pi)^{-2n} \int_{\mathbb{C}^n} (\rho(\zeta)f)(x)\left(\int_{\mathbb{C}^n} \hat{\sigma}(\zeta - w)\hat{\tau}(w)e^{\frac{1}{4}i[\zeta - w, w]}dw\right) d\zeta.$$

Since

$$[w, w] = 0,$$

it follows that we can let ω be the function on $\mathbb{R}^n \times \mathbb{R}^n$ such that

$$\hat{\omega}(\zeta) = (2\pi)^{-n} \int_{\mathbb{C}^n} \hat{\sigma}(\zeta - w)\hat{\tau}(w)e^{\frac{1}{4}i[\zeta, w]}dw, \quad \zeta \in \mathbb{C}^n.$$

Thus,

$$\hat{\omega} = (2\pi)^{-n}(\hat{\sigma} *_{1/4} \hat{\tau}).$$

$\square$

Historical Notes

Eugene Wigner (1902–1995) was a Hungarian-American theoretical physicist and mathematician. He was one of the three winners of the Nobel Prize in Physics in 1963. The Wigner transform introduced in this chapter is a mild modification of the Wigner transform of a function f in $L^2(\mathbb{R})$ used by him in [53] as a substitute for the joint probability distribution of position and momentum in the state f. His famous statement "The Unreasonable Effectiveness of Mathematics in Natural Sciences" is well known among many mathematicians.

Hermann Klaus Hugo Weyl (1885–1955) was a German mathematician and theoretical physicist. His works permeate much of mathematics. Of particular interest in this book is his treatise titled *Gruppentheorie und Quantenmechanik* translated by H. P. Robertson and reprinted by Dover [52] in 1950. In it is contained ideas of quantizations, which are now called Weyl transforms. The connections between pseudo-differential operators and Weyl transforms were explained in the 1960s.

José Enrique Moyal (1910–1998) was a mathematical physicist born in Jerusalem and passed away in Canberra, Australia. He is best remembered for his works in quantum mechanics and statistical mechanics [32].

Exercises

1. Prove that for all points q and p in $\mathbb{R}^n$, $\rho(q,p) : L^2(\mathbb{R}^n) \to L^2(\mathbb{R}^n)$ is a unitary operator.

2. Prove Proposition 20.3.

3. Prove the Moyal identity for the Fourier–Wigner transform, *i.e.*,

$$(V(f_1, g_1), V(f_2, g_2)) = (f_1, f_2)\overline{(g_1, g_2)}$$

   for all functions f_1, g_1, f_2, and g_2 in $L^2(\mathbb{R}^n)$.

4. Let f and g be functions in $L^2(\mathbb{C}^n)$ and let λ be a nonzero real number. Prove that

$$f *_\lambda g = g *_{-\lambda} f.$$

5. Let f and g be functions in $L^2(\mathbb{C}^n)$ and let λ be a nonzero real number. Prove that

$$h = \tilde{f} *_\lambda \tilde{g} \Rightarrow \tilde{h} = f *_\lambda g. \tag{20.2}$$

6. Let $\lambda \in \mathbb{R} \setminus \{0\}$. Prove that

$$f *_\lambda g \in L^2(\mathbb{C}^n)$$

   for all functions f and g in $L^2(\mathbb{C}^n)$. What happens if $\lambda = 0$?

Chapter 21

Spectral Analysis of Twisted Laplacians

A fundamental ingredient in the construction of the heat kernel of L_τ for $\tau \in \mathbb{R} \setminus \{0\}$ is the spectral analysis of L_τ, which we now present. The contents in this chapter and the constructions of the heat kernel and Green function for L_τ, $\tau \in \mathbb{R} \setminus \{0\}$, in the following two chapters are the τ-versions of those in the paper [55].

Let f and g be functions in the Schwartz space $\mathcal{S}$. Then for $\tau \in \mathbb{R} \setminus \{0\}$, the τ-Fourier–Wigner transform $V_\tau(f, g)$ of f and g is defined by

$$V_\tau(f, g)(q, p) = (2\pi)^{-1/2} |\tau|^{1/2} \int_{-\infty}^{\infty} e^{i\tau qy} f\left(y + \frac{p}{2}\right) \overline{g\left(y - \frac{p}{2}\right)} dy$$

for all q and p in $\mathbb{R}$.

For $\tau \in \mathbb{R} \setminus \{0\}$ and $k = 0, 1, 2, \ldots$, we define the function e_k^τ on $\mathbb{R}$ by

$$e_k^\tau(x) = |\tau|^{1/4} e_k(\sqrt{|\tau|} x), \quad x \in \mathbb{R},$$

where e_k is the Hermite function defined in (14.1). For $j, k = 0, 1, 2, \ldots$, we define the function $e_{j,k}^\tau$ on $\mathbb{R}^2$ by

$$e_{j,k}^\tau = V_\tau(e_j^\tau, e_k^\tau).$$

We denote $e_{j,k}^1$ by $e_{j,k}$ and we first establish the connection of $\{e_{j,k}^\tau : j, k = 0, 1, 2, \ldots\}$ with $\{e_{j,k} : j, k = 0, 1, 2, \ldots\}$.

Theorem 21.1 *Let $\tau \in \mathbb{R} \setminus \{0\}$. Then for $j, k = 0, 1, 2, \ldots$,*

$$e_{j,k}^\tau(q, p) = |\tau|^{1/2} e_{j,k}\left(\frac{\tau}{\sqrt{|\tau|}} q, \sqrt{|\tau|} p\right), \quad q, p \in \mathbb{R}.$$

DOI: 10.1201/9781003206781-21

Proof Let $\tau \in \mathbb{R} \setminus \{0\}$. Then using the Fourier–Wigner transform and a change of variables, we get

$$
\begin{aligned}
&e_{j,k}^{\tau}(q,p) \\
&= V_{\tau}(e_j^{\tau}, e_k^{\tau})(q,p) \\
&= (2\pi)^{-1/2} |\tau|^{1/2} \int_{-\infty}^{\infty} e^{i\tau qy} e_j^{\tau}\left(y + \frac{p}{2}\right) \overline{e_k^{\tau}\left(y - \frac{p}{2}\right)} dy \\
&= (2\pi)^{-1/2} |\tau| \int_{-\infty}^{\infty} e^{i\tau qy} e_j\left(\sqrt{|\tau|}\left(y + \frac{p}{2}\right)\right) \overline{e_k\left(\sqrt{|\tau|}\left(y - \frac{p}{2}\right)\right)} dy \\
&= (2\pi)^{-1/2} |\tau|^{1/2} \int_{-\infty}^{\infty} e^{i\tau qy/\sqrt{|\tau|}} e_j\left(y + \frac{\sqrt{|\tau|}p}{2}\right) \overline{e_k\left(y - \frac{\sqrt{|\tau|}p}{2}\right)} dy \\
&= |\tau|^{1/2} e_{j,k}\left(\frac{\tau}{\sqrt{|\tau|}}q, \sqrt{|\tau|}p\right)
\end{aligned}
$$

for all q and p in $\mathbb{R}$. $\qquad\square$

Remark 21.2 Theorem 21.1 suggests that in order to work with the functions $e_{j,k}^{\tau}$ for $\tau \in \mathbb{R} \setminus \{0\}$ and $j,k = 0,1,2,\dots$, it is best to begin with $e_{j,k}$.

Theorem 21.3 $\{e_{j,k} : j,k = 0,1,2,\dots\}$ *is an orthonormal basis for* $L^2(\mathbb{C})$.

Proof Using the Moyal identity for the Fourier–Wigner transform and the Plancherel theorem, we get for all nonnegative integers j_1, j_2, k_1, and k_2,

$$
(e_{j_1,k_1}, e_{j_2,k_2}) = (V(e_{j_1}, e_{k_1}), V(e_{j_2}, e_{k_2})) = (e_{j_1}, e_{j_2})(e_{k_1}, e_{k_2}) = 0
$$

unless $j_1 = j_2$ and $k_1 = k_2$; and if $j_1 = j_2$ and $k_1 = k_2$, we have

$$
(e_{j_1,k_1}, e_{j_2,k_2}) = 1.
$$

So, $\{e_{j,k} : j,k = 0,1,2,\dots\}$ is an orthonormal set in $L^2(\mathbb{R}^2)$. To complete the proof, we have to show that if $f \in L^2(\mathbb{R}^2)$ is such that

$$
(f, e_{j,k}) = 0
$$

for $j,k = 0,1,2,\dots$, then $f = 0$ almost everywhere on $\mathbb{R}^2$. To this end, let $g \in L^2(\mathbb{R}^2)$ be such that $\hat{g} = \overline{f}$. Then using the definition of the Weyl transform and the adjoint formula for the Fourier transform, we get

$$
\begin{aligned}
(W_g e_j, e_k) &= (2\pi)^{-1/2} \int_{-\infty}^{\infty} \int_{-\infty}^{\infty} g(x,\xi) W(e_j, e_k)(x,\xi) \, dx \, d\xi \\
&= (2\pi)^{-1/2} \int_{-\infty}^{\infty} \int_{-\infty}^{\infty} \overline{f(q,p)} e_{j,k}(q,p) \, dq \, dp = 0
\end{aligned}
$$

for $j,k = 0,1,2,\dots$. So,

$$
W_g e_j = 0, \quad j = 0,1,2,\dots.
$$

Now, let $h \in L^2(\mathbb{R})$ and let ε be any positive number. Since $\{e_j : j = 0, 1, 2, \dots\}$ is an orthonormal basis for $L^2(\mathbb{R})$, we can find a finite linear combination $\sum_{l=1}^{N} a_{j_l} e_{j_l}$ such that

$$\left\| \sum_{l=1}^{N} a_{j_l} e_{j_l} - h \right\|_2 < \varepsilon.$$

So, by Exercise 5 in this chapter,

$$\|W_g h\|_2 \leq \left\| W_g \left(h - \sum_{l=1}^{N} a_{j_l} e_{j_l} \right) \right\|_2 + \left\| W_g \left(\sum_{l=1}^{N} a_{j_l} e_{j_l} \right) \right\|_2 \leq \varepsilon \|W_g\|_*,$$

where $\|W_g\|_*$ is the operator norm of W_g. Since ε and h are arbitrary, it follows that

$$W_g h = 0, \quad h \in L^2(\mathbb{R}).$$

But then for all h in $L^2(\mathbb{R})$, we get by (20.1)

$$
\begin{aligned}
(W_g h)(x) &= (2\pi)^{-1} \int_{-\infty}^{\infty} \int_{-\infty}^{\infty} \hat{g}(q, p)(\rho(q, p)h)(x) \, dq \, dp \\
&= (2\pi)^{-1} \int_{-\infty}^{\infty} \int_{-\infty}^{\infty} \hat{g}(q, p) e^{iqx + \frac{1}{2}iqp} h(x + p) \, dq \, dp = 0, \quad x \in \mathbb{R}.
\end{aligned}
$$

Let us change the variable from p to p' by means of the formula $p' = x + p$. Then we get

$$(W_g h)(x) = (2\pi)^{-1} \int_{-\infty}^{\infty} h(p') \left(\int_{-\infty}^{\infty} \hat{g}(q, p' - x) e^{iqx + \frac{1}{2}iq(p'-x)} dq \right) dp' = 0$$

for all x in $\mathbb{R}$. Thus, for almost all x and p' in $\mathbb{R}$,

$$(2\pi)^{-1/2} \int_{-\infty}^{\infty} \hat{g}(q, p' - x) e^{iqx + \frac{1}{2}iq(p'-x)} dq = 0.$$

So, by the Fourier inversion formula, we have for almost all x and p' in $\mathbb{R}$,

$$(\mathcal{F}_2 g) \left(\frac{1}{2}p' + \frac{1}{2}x, p' - x \right) = 0,$$

where $\mathcal{F}_2$ denotes the Fourier transform with respect to the second variable. Therefore $g = 0$ and the proof is complete. $\qquad \square$

The spectral analysis of L depends on the following result, which is an analog of Theorem 14.2 for the operators Z and $\overline{Z}$.

Theorem 21.4 *For all z in $\mathbb{C}$,*

$$(Z e_{j,k})(z) = i(2k)^{1/2} e_{j,k-1}(z), \quad j = 0, 1, 2, \dots, \quad k = 1, 2, \dots,$$

and

$$(\overline{Z} e_{j,k})(z) = i(2k + 2)^{1/2} e_{j,k+1}(z), \quad j, k = 0, 1, 2, \dots.$$

We give the complete spectral analysis of the twisted Laplacian L in the following theorem.

Theorem 21.5 *For all* $j, k = 0, 1, 2, \dots,$

$$Le_{j,k} = (2k+1)e_{j,k}.$$

Theorem 21.5 says that for $k = 0, 1, 2, \dots$, the integer $2k+1$ is an eigenvalue of L and $e_{j,k}$ is an eigenfunction of L corresponding to $2k+1$ for $j = 0, 1, 2, \dots$. We note that each eigenvalue of L has infinite multiplicity.

The proofs of Theorems 21.4 and 21.5 are left as exercises.

Theorems 21.3 and 21.5 are the underpinnings of the spectral analysis of the twisted Laplacian L_τ for $\tau \in \mathbb{R} \setminus \{0\}$, which we now present. We begin with the following theorem.

An orthonormal basis for $L^2(\mathbb{C})$ that is closely related to that in Theorem 21.3 is given in the following theorem.

Theorem 21.6 *Let* $\tau \in \mathbb{R} \setminus \{0\}$. *Then* $\{e_{j,k}^\tau : j, k = 0, 1, 2, \dots\}$ *is an orthonormal basis for* $L^2(\mathbb{C})$.

Proof For nonnegative integers α, β, μ, and ν,

$$(e_{\alpha,\beta}^\tau, e_{\mu,\nu}^\tau) = |\tau| \int_{-\infty}^\infty \int_{-\infty}^\infty e_{\alpha,\beta}\left(\frac{\tau}{\sqrt{|\tau|}}q, \sqrt{|\tau|}p\right) \overline{e_{\mu,\nu}\left(\frac{\tau}{\sqrt{|\tau|}}q, \sqrt{|\tau|}p\right)} \, dq \, dp.$$

Let $q' = \dfrac{\tau}{\sqrt{|\tau|}}q$ and $p' = \sqrt{|\tau|}p$. Then

$$(e_{\alpha,\beta}^\tau, e_{\mu,\nu}^\tau) = \int_{-\infty}^\infty \int_{-\infty}^\infty e_{\alpha,\beta}(q',p')\overline{e_{\mu,\nu}(q',p')} \, dq' \, dp'.$$

Thus, by Theorem 21.3, $\{e_{j,k}^\tau : j, k = 0, 1, 2, \dots\}$ is an orthonormal set in $L^2(\mathbb{C})$. Now, let $f \in L^2(\mathbb{C})$ be such that

$$(f, e_{j,k}^\tau) = 0, \quad j, k = 0, 1, 2, \dots.$$

Then

$$\int_{-\infty}^\infty \int_{-\infty}^\infty f(q,p)\overline{e_{j,k}\left(\frac{\tau}{\sqrt{|\tau|}}q, \sqrt{|\tau|}p\right)} \, dq \, dp = 0$$

for $j, k = 0, 1, 2, \dots$. Let q' and p' be as above in this proof. Then

$$\int_{-\infty}^\infty \int_{-\infty}^\infty f\left(\frac{\sqrt{|\tau|}}{\tau}q', \frac{1}{\sqrt{|\tau|}}p'\right) \overline{e_{j,k}(q',p')} \, dq' \, dp' = 0$$

for $j, k = 0, 1, 2, \dots$. By Theorem 21.3, $\{e_{j,k} : j, k = 0, 1, 2, \dots\}$ is complete in $L^2(\mathbb{C})$. Therefore

$$f\left(\frac{\sqrt{|\tau|}}{\tau}q, \frac{1}{\sqrt{|\tau|}}p\right) = 0$$

for almost all (q, p) in $\mathbb{C}$. Hence $f(q, p) = 0$ for almost all (q, p) in $\mathbb{C}$. This completes the proof. □

We can now give the complete spectral analysis of the twisted Laplacian L_τ for $\tau \in \mathbb{R} \setminus \{0\}$.

Theorem 21.7 *Let $\tau \in \mathbb{R} \setminus \{0\}$. Then for $j, k = 0, 1, 2, \ldots,$*

$$L_\tau e_{j,k}^\tau = (2k + 1)|\tau| e_{j,k}^\tau.$$

The proof of Theorem 21.7 is best achieved by first transforming L_τ into L. The following lemma says that in a suitable coordinate system, the twisted Laplacian L_τ, $\tau \in \mathbb{R} \setminus \{0\}$, is just $|\tau| L$.

Lemma 21.8 *Let $x = \frac{\tau}{\sqrt{|\tau|}} q$ and $y = \sqrt{|\tau|} p$. Then*

$$L_\tau^{(q,p)} = |\tau| L^{(x,y)},$$

where

$$L_\tau^{(q,p)} = -\left(\frac{\partial^2}{\partial q^2} + \frac{\partial^2}{\partial p^2} \right) + \frac{1}{4}(q^2 + p^2)\tau^2 - i\left(q\frac{\partial}{\partial p} - p\frac{\partial}{\partial q} \right)\tau$$

and

$$L^{(x,y)} = -\left(\frac{\partial^2}{\partial x^2} + \frac{\partial^2}{\partial y^2} \right) + \frac{1}{4}(x^2 + y^2) - i\left(x\frac{\partial}{\partial y} - y\frac{\partial}{\partial x} \right).$$

Proof We first note that

$$\frac{\partial}{\partial q} = \frac{\partial}{\partial x} \frac{\tau}{\sqrt{|\tau|}}$$

and

$$\frac{\partial}{\partial p} = \frac{\partial}{\partial y} \sqrt{|\tau|}.$$

So,

$$\frac{\partial^2}{\partial q^2} = \frac{\partial^2}{\partial x^2} |\tau|$$

and

$$\frac{\partial^2}{\partial p^2} = \frac{\partial^2}{\partial y^2} |\tau|.$$

Therefore

$$
\begin{aligned}
L_\tau^{(q,p)} &= -\left(\frac{\partial^2}{\partial q^2} + \frac{\partial^2}{\partial p^2} \right) + \frac{1}{4}(q^2 + p^2)\tau^2 - i\left(q\frac{\partial}{\partial p} - p\frac{\partial}{\partial q} \right)\tau \\
&= -|\tau|\left(\frac{\partial^2}{\partial x^2} + \frac{\partial^2}{\partial y^2} \right) + |\tau|\frac{1}{4}(x^2 + y^2) - |\tau| i\left(x\frac{\partial}{\partial y} - y\frac{\partial}{\partial x} \right) \\
&= |\tau| L^{(x,y)}.
\end{aligned}
$$

□

Proof of Theorem 21.7 Using Theorem 21.1, Theorem 21.5, and Lemma 21.8, we get

$$
\begin{aligned}
(L_\tau e^\tau_{j,k})(q,p) &= |\tau|^{1/2}(Le_{j.k})(x,y) \\
&= |\tau|^{1/2}(2k+1)|\tau|e_{j.k}(x,y) \\
&= (2k+1)|\tau|e^\tau_{j,k}(q,p)
\end{aligned}
$$

for all (q,p) in $\mathbb{C}$.   □

Exercises

1. Prove that for $\tau \in \mathbb{R} \setminus \{0\}$,

$$
e^\tau_{k,j}(q,p) = \overline{e^\tau_{j,k}(-q,-p)} = |\tau|^{1/2}\overline{e_{j,k}\left(\frac{\tau}{\sqrt{|\tau|}}q, -\sqrt{|\tau|}p\right)}, \quad q,p \in \mathbb{R}.
$$

2. Prove Theorem 21.4.

3. Prove Theorem 21.5.

4. Let σ and τ be fixed nonzero real numbers. Let f be a measurable function on $\mathbb{R}^2$ such that

$$
f(\sigma q, \tau p) = 0
$$

for almost all (q,p) in $\mathbb{R}^2$. Prove that

$$
f(q,p) = 0
$$

for almost all (q,p) in $\mathbb{R}^2$.

5. Let $\sigma \in L^2(\mathbb{C})$. Prove that $W_\sigma : L^{(}\mathbb{R}) \to L^2(\mathbb{R})$ is a bounded linear operator in the sense that there exists a positive constant C such that

$$
\|W_\sigma f\|_2 \le C\|f\|_2, \quad f \in L^2(\mathbb{R}).
$$

(The least positive constant C for which the preceding inequality holds is denoted by $\|W_\sigma\|_*$.)

Chapter 22

Heat Kernels Related to the Heisenberg Group

We give in this chapter the heat kernel of the twisted Laplacian L_τ, $\tau \in \mathbb{R}\backslash\{0\}$, and the heat kernel of the sub-Laplacian $\mathcal{L}$ on the Heisenberg group $\mathbb{H}^1$. The following formula is the main tool for the construction of the heat kernel of L_τ for $\tau \in \mathbb{R} \backslash \{0\}$.

Theorem 22.1 *For all nonnegative integers α, β, μ, and ν,*

$$e_{\alpha,\beta} *_{1/4} e_{\mu,\nu} = (2\pi)^{1/2}\delta_{\beta,\mu}e_{\alpha,\nu},$$

where $\delta_{\beta,\mu}$ is the Kronecker delta given by

$$\delta_{\beta,\mu} = \begin{cases} 1, & \beta = \mu, \\ 0, & \beta \neq \mu. \end{cases}$$

Proof Let φ and ψ be functions in $\mathcal{S}$. Then by the definition of the Weyl transform, Proposition 20.3, and the Moyal identity in Theorem 20.4,

$$
\begin{aligned}
(W_{\widehat{e_{\alpha,\beta}}}\varphi, \psi) &= (2\pi)^{-1/2}\int_{\mathbb{C}} \widehat{e_{\alpha,\beta}}(z)W(\varphi,\psi)(z)\,dz \\
&= (2\pi)^{-1/2}\int_{\mathbb{C}} \overline{W(e_\beta, e_\alpha)(z)}W(\varphi,\psi)(z)\,dz \\
&= (2\pi)^{-1/2}(W(\varphi,\psi), W(e_\beta, e_\alpha)) \\
&= (2\pi)^{-1/2}(\varphi, e_\beta)\overline{(\psi, e_\alpha)} \\
&= (2\pi)^{-1/2}(\varphi, e_\beta)(e_\alpha, \psi).
\end{aligned}
$$

Hence for all φ in $\mathcal{S}$,

$$W_{\widehat{e_{\alpha,\beta}}}\varphi = (2\pi)^{-1/2}(\varphi, e_\beta)e_\alpha$$

and therefore

$$
\begin{aligned}
W_{\widehat{e_{\alpha,\beta}}}W_{\widehat{e_{\mu,\nu}}}\varphi &= (2\pi)^{-1/2}(W_{\widehat{e_{\mu,\nu}}}\varphi, e_\beta)e_\alpha \\
&= (2\pi)^{-1}(\varphi, e_\nu)(e_\mu, e_\beta)e_\alpha \\
&= (2\pi)^{-1}(\varphi, e_\nu)\delta_{\beta,\mu}e_\alpha \\
&= (2\pi)^{-1/2}W_{\delta_{\beta,\mu}\widehat{e_{\alpha,\nu}}}\varphi. \quad\quad (22.1)
\end{aligned}
$$

DOI: 10.1201/9781003206781-22

By Theorem 20.6 on the product of two Weyl transforms and the Fourier inversion formula in, say, Theorem 5.14, we have

$$W_{\widehat{e_{\alpha,\beta}}} W_{\widehat{e_{\mu,\nu}}} = W_{\omega},$$

where

$$\hat{\omega} = (2\pi)^{-1}(\widehat{\widehat{e_{\alpha,\beta}}} *_{1/4} \widehat{\widehat{e_{\mu,\nu}}}) = (2\pi)^{-1}(\widehat{e_{\alpha,\beta}} *_{1/4} \widehat{e_{\mu,\nu}}). \tag{22.2}$$

By (22.1),

$$\omega = (2\pi)^{-1/2}\delta_{\beta,\mu}\widehat{e_{\alpha,\nu}}$$

and hence the Fourier inversion formula gives

$$\hat{\omega} = (2\pi)^{-1/2}\delta_{\beta,\mu}\widehat{\widehat{e_{\alpha,\nu}}} = (2\pi)^{-1/2}\delta_{\beta,\mu}\widetilde{e_{\alpha,\nu}}. \tag{22.3}$$

So, by (20.2), (22.2), and (22.3),

$$\delta_{\beta,\mu}e_{\alpha,\nu} = (2\pi)^{-1/2}(e_{\alpha,\beta} *_{1/4} e_{\mu,\nu})$$

and the proof is complete. □

The following theorem is a τ-version of Theorem 22.1.

Theorem 22.2 *Let $\tau \in \mathbb{R} \setminus \{0\}$. Then for all nonnegative integers α, β, μ, and ν,*

$$e^{\tau}_{\alpha,\beta} *_{\tau/4} e^{\tau}_{\mu,\nu} = (2\pi)^{1/2}|\tau|^{-1/2}\delta_{\beta,\mu}e^{\tau}_{\alpha,\nu},$$

where $\delta_{\beta,\mu}$ is the Kronecker delta in the preceding theorem.

Proof Let $z = (q,p)$ and $w = (x,\xi)$ be points in $\mathbb{C}$. Then by Theorem 21.1, the definition of a twisted convolution and most importantly, Theorem 22.1,

$$(e^{\tau}_{\alpha,\beta} *_{\tau/4} e^{\tau}_{\mu,\nu})(z)$$

$$= \int_{-\infty}^{\infty}\int_{-\infty}^{\infty} e^{\tau}_{\alpha,\beta}(z-w)e^{\tau}_{\mu,\nu}(w)e^{i\frac{\tau}{4}[z,w]}dw$$

$$= |\tau|\int_{-\infty}^{\infty}\int_{-\infty}^{\infty} e_{\alpha,\beta}\left(\frac{\tau}{\sqrt{|\tau|}}(q-x), \sqrt{|\tau|}(p-\xi)\right) \times$$

$$\times e_{\mu,\nu}\left(\frac{\tau}{\sqrt{|\tau|}}x, \sqrt{|\tau|}\xi\right)e^{i\frac{\tau}{2}(xp-\xi q)}dx\,d\xi.$$

Let

$$q' = \frac{\tau}{\sqrt{|\tau|}}x$$

and

$$p' = \sqrt{|\tau|}\xi.$$

Then

$$(e^\tau_{\alpha,\beta} *_{\tau/4} e^\tau_{\mu,\nu})(z)$$

$$= \int_{-\infty}^{\infty} \int_{-\infty}^{\infty} e_{\alpha,\beta}\left(\frac{\tau}{\sqrt{|\tau|}}q - q', \sqrt{|\tau|}p - p'\right) \times$$

$$\times e_{\mu,\nu}(q',p')e^{i\frac{\tau}{2}\left(\frac{\sqrt{|\tau|}}{\tau}q'p - \frac{1}{\sqrt{|\tau|}}p'q\right)} dq'\, dp'$$

$$= |\tau|^{-1/2}|\tau|^{1/2}(e_{\alpha,\beta} *_{1/4} e_{\mu,\nu})\left(\frac{\tau}{\sqrt{|\tau|}}q, \sqrt{|\tau|}p\right)$$

$$= (2\pi)^{1/2}|\tau|^{-1/2}\delta_{\beta,\mu}|\tau|^{1/2}e_{\alpha,\nu}\left(\frac{\tau}{\sqrt{|\tau|}}q, \sqrt{|\tau|}p\right)$$

$$= (2\pi)^{1/2}|\tau|^{-1/2}\delta_{\beta,\mu}e^\tau_{\alpha,\nu}(q,p),$$

as asserted. $\qquad\qquad\qquad\qquad\qquad\qquad\qquad\qquad\qquad\qquad\qquad$ $\square$

We also need the following version of Mehler's formula.

Theorem 22.3 *For all $z = q + ip \in \mathbb{C}$ and all $r \in (-1,1)$,*

$$\sum_{k=0}^{\infty} e_{k,k}(z)r^k = (2\pi)^{-1/2}\frac{1}{1-r}e^{-\frac{1}{4}|z|^2\frac{1+r}{1-r}}$$

and the convergence of the series is absolute and uniform on compact subsets of $(-1,1)$.

Proof For all y and p in $\mathbb{R}$, and all $r \in (-1,1)$, we get by Theorem 14.7

$$\sum_{k=0}^{\infty} e_k\left(y + \frac{p}{2}\right) e_k\left(y - \frac{p}{2}\right) r^k$$

$$= \frac{1}{\sqrt{\pi}}(1 - r^2)^{-1/2}e^{-\frac{1}{2}\frac{1+r^2}{1-r^2}\left(2y^2 + \frac{p^2}{2}\right) + \frac{2r}{1-r^2}\left(y^2 - \frac{p^2}{4}\right)}$$

$$= \frac{1}{\sqrt{\pi}}(1 - r^2)^{-1/2}e^{-\frac{1+r^2}{1-r^2}y^2 - \frac{1+r^2}{1-r^2}\frac{p^2}{4} + \frac{2r}{1-r^2}y^2 - \frac{2r}{1-r^2}\frac{p^2}{4}}$$

$$= \frac{1}{\sqrt{\pi}}(1 - r^2)^{-1/2}e^{-\frac{1-2r+r^2}{1-r^2}y^2 - \frac{1+2r+r^2}{1-r^2}\frac{p^2}{4}}$$

$$= \frac{1}{\sqrt{\pi}}(1 - r^2)^{-1/2}e^{-\frac{1-r}{1+r}y^2 - \frac{1+r}{1-r}\frac{p^2}{4}}.$$

Taking the inverse Fourier transform of the preceding formula with respect to y and using Proposition 4.6, we obtain

$$
\begin{aligned}
&\sum_{k=0}^{\infty} e_{k,k}(z) r^k \\
&= \frac{1}{\sqrt{\pi}} (1-r^2)^{-1/2} e^{-\frac{1+r}{1-r}\frac{\rho^2}{4}} (2\pi)^{-1/2} \int_{-\infty}^{\infty} e^{iqy} e^{-\frac{1-r}{1+r}y^2}\, dy \\
&= (2\pi)^{-1/2} \frac{1}{1-r} e^{-\frac{1}{4}|z|^2 \frac{1+r}{1-r}}
\end{aligned}
$$

and the proof is complete. □

Let $\tau \in \mathbb{R} \setminus \{0\}$. Using Theorem 21.7 and the series expansions in Hilbert spaces, we get for all functions f in $L^2(\mathbb{C})$,

$$
e^{-\rho L_\tau} f = \sum_{k=0}^{\infty} \sum_{j=0}^{\infty} e^{-(2k+1)|\tau|\rho} (f, e_{j,k}^\tau) e_{j,k}^\tau, \quad \rho > 0.
$$

So, for $\rho > 0$,

$$
e^{-\rho L_\tau} f = \sum_{k=0}^{\infty} e^{-(2k+1)|\tau|\rho} \sum_{j=0}^{\infty} (f, e_{j,k}^\tau) e_{j,k}^\tau
$$

and our first task is to compute $\sum_{j=0}^{\infty} (f, e_{j,k}^\tau) e_{j,k}^\tau$. To this end, we note that for $k = 0, 1, 2, \ldots,$

$$
\begin{aligned}
f *_{\tau/4} e_{k,k}^\tau &= \sum_{j=0}^{\infty} \sum_{l=0}^{\infty} (f, e_{j,l}^\tau)(e_{j,l}^\tau *_{\tau/4} e_{k,k}^\tau) \\
&= \sum_{j=0}^{\infty} \sum_{l=0}^{\infty} (f, e_{j,l}^\tau)(2\pi)^{1/2}|\tau|^{-1/2}\delta_{l,k} e_{j,k}^\tau \\
&= (2\pi)^{1/2}|\tau|^{-1/2} \sum_{j=0}^{\infty} (f, e_{j,k}^\tau) e_{j,k}^\tau.
\end{aligned}
$$

Hence for $k = 0, 1, 2, \ldots,$

$$
\sum_{j=0}^{\infty} (f, e_{j,k}^\tau) e_{j,k}^\tau = (2\pi)^{-1/2}|\tau|^{1/2}(f *_{\tau/4} e_{k,k}^\tau).
$$

Therefore

$$
e^{-\rho L_\tau} f = (2\pi)^{-1/2}|\tau|^{1/2} \sum_{k=0}^{\infty} e^{-(2k+1)|\tau|\rho}(e_{k,k}^\tau *_{-\tau/4} f), \quad \rho > 0.
$$

Now, using Theorem 21.1 and Mehler's formula in Theorem 22.3, we get for all $z = q + ip$ in $\mathbb{C}$ and $\rho > 0$,

$$(2\pi)^{-1/2}|\tau|^{1/2} \sum_{k=0}^{\infty} e^{-(2k+1)|\tau|\rho} e_{k,k}^{\tau}(q,p)$$

$$= (2\pi)^{-1/2}|\tau|e^{-|\tau|\rho} \sum_{k=0}^{\infty} e^{-2k|\tau|\rho} e_{k,k} \left(\frac{\tau}{\sqrt{|\tau|}} q, \sqrt{|\tau|} p \right)$$

$$= (2\pi)^{-1}|\tau|e^{-|\tau|\rho} \frac{1}{1 - e^{-2|\tau|\rho}} e^{-|\tau||z|^2 \frac{1}{4} \frac{1+e^{-2|\tau|\rho}}{1-e^{-2|\tau|\rho}}}$$

$$= \frac{1}{4\pi} \frac{\tau}{\sinh(\tau\rho)} e^{-\frac{1}{4}\tau|z|^2 \coth(\tau\rho)}.$$

So, the heat kernel κ_ρ^{τ}, $\rho > 0$, of L_τ is given by

$$\kappa_\rho^{\tau}(z,w) = \frac{1}{4\pi} \frac{\tau}{\sinh(\tau\rho)} e^{-\frac{1}{4}\tau|z-w|^2 \coth(\tau\rho)} e^{-i\frac{1}{4}\tau[z,w]}, \quad z,w \in \mathbb{C}.$$

Hence by (19.3), we have the following result.

Theorem 22.4 *Let $\tau \in \mathbb{R} \setminus \{0\}$. Then for $\rho > 0$,*

$$K_\rho^{\tau}(z) = (2\pi)^{-1/2} \frac{1}{4\pi} \frac{\tau}{\sinh(\tau\rho)} e^{-\frac{1}{4}\tau|z|^2 \coth(\tau\rho)}, \quad z \in \mathbb{C}.$$

An immediate application of Theorem 22.4 is the following formula for the heat kernel K_ρ of the sub-Laplacian $\mathcal{L}$.

Theorem 22.5 *For $\rho > 0$, the heat kernel K_ρ of the sub-Laplacian $\mathcal{L}$ is given by*

$$K_\rho(z,t) = \frac{1}{8\pi^2} \int_{-\infty}^{\infty} e^{-it\tau} \frac{\tau}{\sinh(\tau\rho)} e^{-\frac{1}{4}\tau|z|^2 \coth(\tau\rho)} d\tau, \quad (z,t) \in \mathbb{H}^1.$$

The heat kernel of the sub-Laplacian derived in this chapter can be obtained by different techniques in, for instance, [13, 27].

Historical Notes

Leopold Kronecker (1823–1891), a German mathematician, worked in algebra and number theory. He saw the connection of the elliptic modular function with algebraic numbers. The Kronecker delta resonates well with unit impulse in digital signal processing.

Exercises

1. Prove Theorem 22.5.

2. Let f be the function on $\mathbb{H}^1 \times (\mathbb{R} \setminus \{0\})$ defined by

$$f(z,t,\tau) = \tau\left(it + \frac{1}{4}|z|^2\coth\tau\right), \quad (z,t,\tau) \in \mathbb{H}^1 \times (\mathbb{R} \setminus \{0\}),$$

and let V be the function defined on $\mathbb{R} \setminus \{0\}$ by

$$V(\tau) = \frac{\tau}{\sinh\tau}, \quad \tau \in \mathbb{R} \setminus \{0\}.$$

Prove that

$$K_\rho(z,t) = \frac{1}{8\pi^2\rho^2} \int_{-\infty}^{\infty} e^{-f(z,t,\tau)/\rho} V(\tau)\, d\tau, \quad (z,t) \in \mathbb{H}^1.$$

3. Let $\Delta_{\mathbb{H}^1}$ be the Heisenberg Laplacian *i.e.*, the Laplacian on the Heisenberg group $\mathbb{H}^1$ defined by

$$\Delta_{\mathbb{H}^1} = -(X^2 + Y^2 + T^2).$$

Find a formula for the heat kernel $K_\rho^{\Delta_{\mathbb{H}^1}}$, $\rho > 0$, of $\Delta_{\mathbb{H}^1}$.

Chapter 23

Green Functions Related to the Heisenberg Group

The Green function G_τ of L_τ for $\tau \in \mathbb{R} \setminus \{0\}$ is related to the heat kernel κ_ρ, $\rho > 0$, by the formula

$$G_\tau(z, w) = \int_0^\infty \kappa_\rho^\tau(z, w) \, d\rho$$

for all z and w in $\mathbb{C}$. The explicit formula for G_τ is given by the following theorem.

Theorem 23.1 *Let $\tau \in \mathbb{R} \setminus \{0\}$. Then for all z and w in $\mathbb{C}$,*

$$G_\tau(z, w) = e^{-i\frac{1}{4}\tau[z,w]}\frac{1}{4\pi}K_0\left(\frac{1}{4}|\tau|\,|z-w|^2\right),$$

where K_0 is the modified Bessel function of order 0 given by

$$K_0(x) = \int_0^\infty e^{-x\cosh t}dt, \quad x > 0.$$

Proof Let z and w be in $\mathbb{C}$. Then

$$
\begin{aligned}
G_\tau(z, w) &= \int_0^\infty \kappa_\rho^\tau(z, w)\, d\rho \\
&= e^{-i\frac{1}{4}\tau[z,w]}\frac{1}{4\pi}\int_0^\infty \frac{|\tau|}{\sinh(|\tau|\rho)}e^{-\frac{1}{4}|\tau|\,|z-w|^2\coth(|\tau|\rho)}d\rho.
\end{aligned}
$$

Changing the variable of integration from ρ to v by means of the formula $v = \coth(|\tau|\rho)$, we get

$$dv = -\frac{|\tau|}{\sinh^2(|\tau|\rho)}d\rho.$$

Since

$$\coth^2(|\tau|\rho) = 1 + \operatorname{csch}^2(|\tau|\rho),$$

it follows that

$$G_\tau(z, w) = e^{-i\frac{1}{4}\tau[z,w]}\frac{1}{4\pi}\int_1^\infty \frac{1}{(v^2-1)^{1/2}}e^{-\frac{1}{4}|\tau|\,|z-w|^2 v}dv.$$

DOI: 10.1201/9781003206781-23

Using the formula on page 250 of [31] to the effect that

$$\int_1^\infty (v^2 - 1)^{\gamma - 1} e^{-\mu v} dv = \frac{1}{\sqrt{\pi}} \left(\frac{2}{\mu}\right)^{\gamma - \frac{1}{2}} \Gamma(\gamma) K_{\gamma - \frac{1}{2}}(\mu),$$

where K_ν is the modified Bessel function of order ν given by

$$K_\nu(x) = \int_0^\infty e^{-x \cosh t} \cosh(\nu t)\, dt, \quad x > 0,$$

$\gamma = \frac{1}{2}$, and $\mu = \frac{1}{4}|\tau|\,|z - w|^2$, the asserted formula is proved. $\qquad\square$

To obtain the formula for the Green function of $\mathcal{L}$, we use (19.4) and Theorem 23.1 to obtain

$$
\begin{aligned}
\mathcal{G}^\tau(z) &= (2\pi)^{-1/2} \frac{1}{4\pi} \int_0^\infty \frac{|\tau|}{\sinh(|\tau|\rho)} e^{-\frac{1}{4}|\tau|\,|z|^2 \coth(|\tau|\rho)} d\rho \\
&= (2\pi)^{-1/2} \frac{1}{4\pi} K_0\left(\frac{1}{4}|\tau|\,|z|^2\right)
\end{aligned}
$$

for all z in $\mathbb{C}$. Therefore for all $(z, t) \in \mathbb{H}^1$,

$$
\begin{aligned}
\mathcal{G}(z, t) &= \frac{1}{8\pi^2} \int_{-\infty}^\infty e^{-it\tau} K_0\left(\frac{1}{4}|\tau|\,|z|^2\right) d\tau \\
&= \frac{1}{8\pi^2} \int_{-\infty}^\infty e^{-it\tau} \left(\int_0^\infty e^{-\frac{1}{4}|\tau|\,|z|^2 \cosh \delta} d\delta\right) d\tau \\
&= \frac{1}{8\pi^2} \int_0^\infty \left(\int_{-\infty}^\infty e^{-it\tau} e^{-\frac{1}{4}|\tau|\,|z|^2 \cosh \delta} d\tau\right) d\delta.
\end{aligned}
$$

For all $(z, t) \in \mathbb{H}^1$ and $\delta \in (0, \infty)$,

$$
\begin{aligned}
&\int_{-\infty}^\infty e^{-it\tau} e^{-\frac{1}{4}|\tau|\,|z|^2 \cosh \delta} d\tau \\
&= \int_{-\infty}^0 e^{-it\tau} e^{\frac{1}{4}\tau |z|^2 \cosh \delta} d\tau + \int_0^\infty e^{-it\tau} e^{-\frac{1}{4}\tau |z|^2 \cosh \delta} d\tau \\
&= \left.\frac{e^{\tau\left(\frac{1}{4}|z|^2 \cosh \delta - it\right)}}{\frac{1}{4}|z|^2 \cosh \delta - it}\right|_{-\infty}^0 + \left.\frac{e^{-\tau\left(\frac{1}{4}|z|^2 \cosh \delta + it\right)}}{\frac{1}{4}|z|^2 \cosh \delta + it}\right|_0^\infty \\
&= \frac{\frac{1}{2}|z|^2 \cosh \delta}{(|z|^4/16)\cosh^2 \delta + t^2}.
\end{aligned}
$$

So, for all $(z, t) \in \mathbb{H}^1$,

$$
\begin{aligned}
\mathcal{G}(z, t) &= \frac{1}{8\pi^2} \int_0^\infty \frac{\frac{1}{2}|z|^2 \cosh \delta}{(|z|^4/16)\cosh^2 \delta + t^2} d\delta \\
&= \frac{|z|^2}{16\pi^2} \int_0^\infty \frac{\cosh \delta}{(|z|^4/16)\cosh^2 \delta + t^2} d\delta \\
&= \frac{1}{\pi^2 |z|^2} \int_0^\infty \frac{\cosh \delta}{\cosh^2 \delta + (16t^2/|z|^4)} d\delta.
\end{aligned}
$$

Let $\phi = \sinh \delta$. Then for all $(z,t) \in \mathbb{H}^1$,

$$
\begin{aligned}
\mathcal{G}(z,t) &= \frac{1}{\pi^2 |z|^2} \int_0^\infty \frac{1}{\phi^2 + 1 + (16t^2/|z|^4)} d\phi \\
&= \frac{1}{2\pi^2 |z|^2} \int_{-\infty}^\infty \frac{1}{\phi^2 + 1 + (16t^2/|z|^4)} d\phi \\
&= \frac{1}{2\pi^2 |z|^2} \frac{1}{\sqrt{1+(16t^2/|z|^4)}} \tan^{-1} \frac{\phi}{\sqrt{1+(16t^2/|z|^4)}} \Big|_{-\infty}^\infty \\
&= \frac{1}{2\pi |z|^2} \frac{1}{\sqrt{1+(16t^2/|z|^4)}} \\
&= \frac{1}{2\pi} \frac{1}{\sqrt{|z|^4 + 16t^2}}.
\end{aligned}
$$

The Green function of the sub-Laplacian $\mathcal{L}$ on $\mathbb{H}^1$ should be compared with the Newtonian potential of the Laplacian $-\Delta$ on $\mathbb{R}^3$ given in Chapter 8. The former is

$$
\frac{1}{2\pi} \frac{1}{\sqrt{|z|^4 + 16t^2}}, \quad (z,t) \in \mathbb{H}^1,
$$

and, by the formula for N_3 in Chapter 8, the latter is

$$
\frac{1}{4\pi} \frac{1}{\sqrt{|z|^2 + t^2}}, \quad (z,t) \in \mathbb{R}^3.
$$

It is well known that $\sqrt{|z|^2 + t^2}$ is the distance of point (z,t) from the origin in $\mathbb{R}^3$. It should be pointed out that $\sqrt{|z|^4 + 16t^2}$ can be interpreted as the distance of the point (z,t) from the origin in $\mathbb{H}^1$.

The formula for the Green function of the sub-Laplacian obtained in this chapter can be found in [9, 42].

Exercises

1. Let $\tau \in \mathbb{R} \setminus \{0\}$. Give an explicit formula for the solution u in $L^2(\mathbb{C})$ of the partial differential equation

$$
L_\tau u = f
$$

   for every function f in $L^2(\mathbb{C})$.

2. Let f be a Schwartz function on $\mathbb{R}^2$. Find a solution u on $\mathbb{R}^2$ of the fourth-order partial differential equation $L^2 u = f$ on $\mathbb{R}^2$.

3. Give an explicit formula for a solution u of the partial differential equation

$$
\mathcal{L} u = f,
$$

   where f is a Schwartz function on $\mathbb{H}^1$.

4. Find a formula for the Green function of the Laplacian $\Delta_{\mathbb{H}^1}$ on the Heisenberg group $\mathbb{H}^1$.

Chapter 24

Theta Functions and the Riemann Zeta-Function

We are interested in obtaining the holomorphic continuation of the Riemann zeta-function ζ, defined initially on $\{s \in \mathbb{C} : \operatorname{Re} s > 1\}$, to a meromorphic function on $\mathbb{C}$ with only a simple pole at $s = 1$. We use the standard theory of Fourier series given in [57] to obtain first the Poisson summation formula.

Let f be a Schwartz function on $\mathbb{R}$. Then we define the function g on $\mathbb{R}$ by

$$g(x) = \sum_{n=-\infty}^{\infty} f(x + 2n\pi), \quad x \in \mathbb{R}.$$

It is clear that g is a periodic function on $\mathbb{R}$ with period 2π. Using the Fourier series expansion of g, we get

$$g(x) = \sum_{k=-\infty}^{\infty} a_k e^{ikx}, \quad x \in \mathbb{R},$$

where

$$
\begin{aligned}
a_k &= \frac{1}{2\pi} \int_{-\pi}^{\pi} e^{-ikx} g(x)\, dx \\
&= \frac{1}{2\pi} \int_{-\pi}^{\pi} e^{-ikx} \left(\sum_{n=-\infty}^{\infty} f(x + 2n\pi) \right) dx \\
&= \sum_{n=-\infty}^{\infty} \frac{1}{2\pi} \int_{-\pi}^{\pi} e^{-ikx} f(x + 2n\pi)\, dx
\end{aligned}
$$

for all integers k. Let $y = x + 2n\pi$. Then for all integers k,

$$
\begin{aligned}
a_k &= \sum_{n=-\infty}^{\infty} \frac{1}{2\pi} \int_{(2n-1)\pi}^{(2n+1)\pi} e^{-ik(y-2\pi n)} f(y)\, dy \\
&= \sum_{n=-\infty}^{\infty} \frac{1}{2\pi} \int_{(2n-1)\pi}^{(2n+1)\pi} e^{-iky} f(y)\, dy \\
&= (2\pi)^{-1/2} \hat{f}(k).
\end{aligned}
$$

DOI: 10.1201/9781003206781-24

Therefore

$$\sum_{n=-\infty}^{\infty} f(x + 2n\pi) = (2\pi)^{-1/2} \sum_{k=-\infty}^{\infty} \hat{f}(k) e^{ikx}, \quad x \in \mathbb{R}.$$

If we let $x = 0$ in the preceding formula, then we get

$$\sum_{n=-\infty}^{\infty} f(2n\pi) = (2\pi)^{-1/2} \sum_{k=-\infty}^{\infty} \hat{f}(k).$$

We can write it as

$$\sum_{k=-\infty}^{\infty} \hat{f}(k) = (2\pi)^{1/2} \sum_{n=-\infty}^{\infty} f(2n\pi),$$

which is the Poisson summation formula.

There are many theta functions arising in many applications. We begin with the theta function Θ on $(0, \infty)$ given by

$$\Theta(t) = \sum_{n=-\infty}^{\infty} e^{-n^2 \pi t}, \quad t \in (0, \infty).$$

It is convenient to write the theta function Θ as

$$\Theta(t) = 1 + 2\Psi(t), \quad t \in (0, \infty),$$

where

$$\Psi(t) = \sum_{n=1}^{\infty} e^{-n^2 \pi t}, \quad t \in (0, \infty). \tag{24.1}$$

The function Ψ is also known as a theta function. In fact, it is the theta function that is of particular interest to us in the analysis of the Riemann zeta-function. We first give the following reciprocal formula for the theta function Θ.

Theorem 24.1 *For all $t \in (0, \infty)$, we have*

$$\Theta(t) = \sqrt{\frac{1}{t}} \Theta\left(\frac{1}{t}\right).$$

Proof For a fixed $t \in (0, \infty)$, let f be the Schwartz function on $\mathbb{R}$ defined by

$$f(x) = e^{-tx^2/(4\pi)}, \quad x \in \mathbb{R}.$$

Then

$$f(x) = (D_{\sqrt{t/(2\pi)}} \varphi)(x),$$

where

$$\varphi(x) = e^{-x^2/2}, \quad x \in \mathbb{R}.$$

By Propositions 4.5 and 4.6,

$$\hat{f}(\xi) = \sqrt{\frac{2\pi}{t}} e^{-\pi\xi^2/t}, \quad \xi \in \mathbb{R}.$$

Using the Poisson summation formula, we get

$$\sum_{n=-\infty}^{\infty} f(2n\pi) = (2\pi)^{-1/2} \sum_{k=-\infty}^{\infty} \hat{f}(k)$$

and hence

$$\Theta(t) = \sum_{n=-\infty}^{\infty} e^{-n^2\pi t} = \sqrt{\frac{1}{t}} \sum_{k=-\infty}^{\infty} e^{-k^2\pi/t} = \sqrt{\frac{1}{t}} \Theta\left(\frac{1}{t}\right).$$

□

We need the Mellin transform of the theta function Ψ. To recall the Mellin transform, let f be a suitable function on $(0, \infty)$. Then the Mellin transform of f is the holomorphic function Mf on a strip

$$\{s \in \mathbb{C} : a < \operatorname{Re} s < b\}$$

of the complex plane $\mathbb{C}$ given by

$$(Mf)(s) = \int_0^\infty f(x)x^s \frac{dx}{x}, \quad a < \operatorname{Re} s < b.$$

The gamma function Γ is the Mellin transform of the function f on $(0, \infty)$ given by

$$f(x) = e^{-x}, \quad x > 0.$$

Indeed,

$$\Gamma(s) = \int_0^\infty e^{-x} x^s \frac{dx}{x}, \quad \operatorname{Re} s > 0.$$

We can now give the Mellin transform of the theta function Ψ.

Theorem 24.2 *The Mellin transform $M\Psi$ of the function Ψ defined in (24.1) is given by*

$$(M\Psi)(s) = \pi^{-s}\Gamma(s) \sum_{n=1}^{\infty} \frac{1}{n^{2s}}, \quad \operatorname{Re} s > \frac{1}{2}.$$

Proof For $\operatorname{Re} s > \frac{1}{2}$, we get

$$
\begin{aligned}
(M\Psi)(s) &= \int_0^\infty \Psi(t) t^s \frac{dt}{t} \\
&= \int_0^\infty \left(\sum_{n=1}^\infty e^{-n^2\pi t} \right) t^s \frac{dt}{t} \\
&= \sum_{n=1}^\infty \int_0^\infty e^{-n^2\pi t} t^s \frac{dt}{t}.
\end{aligned}
$$

Let $\tau = n^2 \pi t$. Then for $\operatorname{Re} s > \frac{1}{2}$,

$$
\begin{aligned}
(M\Psi)(s) &= \sum_{n=1}^\infty \int_0^\infty e^{-\tau} \left(\frac{\tau}{n^2\pi} \right)^s \frac{d\tau}{\tau} \\
&= \sum_{n=1}^\infty \frac{1}{n^{2s}\pi^s} \int_0^\infty e^{\tau} \tau^s \frac{d\tau}{\tau} \\
&= \pi^{-s} \Gamma(s) \sum_{n=1}^\infty \frac{1}{n^{2s}}.
\end{aligned} \tag{24.2}
$$

$\square$

The Riemann zeta-function ζ on $\{s \in \mathbb{C} : \operatorname{Re} s > 1\}$ is defined by

$$
\zeta(s) = \sum_{n=1}^\infty \frac{1}{n^s}, \quad \operatorname{Re} s > 1.
$$

It is then easy to check that ζ is holomorphic on $\{s \in \mathbb{C} : \operatorname{Re} s > 1\}$. By Theorem 24.2, we have

$$
\pi^{-s/2} \Gamma\left(\frac{s}{2} \right) \zeta(s) = \int_0^\infty \left(\sum_{n=1}^\infty e^{-n^2\pi t} \right) t^{s/2} \frac{dt}{t}, \quad \operatorname{Re} s > 1.
$$

We now write

$$
\pi^{-s/2} \Gamma\left(\frac{s}{2} \right) \zeta(s) = \int_0^\infty \Psi(t) t^{s/2} \frac{dt}{t} = \left(\int_0^1 + \int_1^\infty \right) \Psi(t) t^{s/2} \frac{dt}{t}.
$$

Using the reciprocal formula in Theorem 24.1 and the relationship between Ψ and Θ given in (24.2), we obtain for all t in $(0, \infty)$,

$$
\begin{aligned}
\Psi(t) &= \frac{1}{2}\Theta(t) - \frac{1}{2} \\
&= \frac{1}{2}\sqrt{\frac{1}{t}}\Theta\left(\frac{1}{t}\right) - \frac{1}{2} \\
&= \frac{1}{2}\sqrt{\frac{1}{t}}\left(2\Psi\left(\frac{1}{t}\right) + 1\right) - \frac{1}{2} \\
&= \sqrt{\frac{1}{t}}\Psi\left(\frac{1}{t}\right) + \frac{1}{2}t^{-1/2} - \frac{1}{2}.
\end{aligned}
$$

Thus,

$$
\begin{aligned}
& \int_0^1 \Psi(t)t^{s/2}\frac{dt}{t} \\
&= \int_0^1 \Psi\left(\frac{1}{t}\right)t^{(s/2)-(1/2)}\frac{dt}{t} + \frac{1}{2}\int_0^1 t^{(s/2)-(1/2)}\frac{dt}{t} - \frac{1}{2}\int_0^1 t^{s/2}\frac{dt}{t} \\
&= \left(\frac{t^{(s/2)-(1/2)}}{s-1} - \frac{t^{s/2}}{s}\right)\Big|_0^1 + \int_0^1 \Psi\left(\frac{1}{t}\right)t^{(s/2)-(1/2)}\frac{dt}{t} \\
&= \frac{1}{s-1} - \frac{1}{s} + \int_0^1 \Psi\left(\frac{1}{t}\right)t^{(s/2)-(1/2)}\frac{dt}{t}.
\end{aligned}
$$

For the integral in the preceding line, let $\tau = \frac{1}{t}$. Then

$$
\int_0^1 \Psi(t)t^{s/2}\frac{dt}{t} = \frac{1}{s-1} - \frac{1}{s} + \int_1^\infty \Psi(\tau)\tau^{(1/2)-(s/2)}\frac{d\tau}{\tau}.
$$

Therefore

$$
\begin{aligned}
& \pi^{-s/2}\Gamma\left(\frac{s}{2}\right)\zeta(s) \\
&= \left(\int_0^1 + \int_1^\infty\right)\Psi(t)t^{s/2}\frac{dt}{t} \\
&= \frac{1}{s-1} - \frac{1}{s} + \int_1^\infty \Psi(t)(t^{s/2} + t^{(1-s)/2})\frac{dt}{t}.
\end{aligned} \tag{24.3}
$$

Since for all $t \in (0, \infty)$,

$$
e^{\pi t/2}\sum_{n=1}^\infty e^{-n^2\pi t} = \sum_{n=1}^\infty e^{-(n^2-(1/2))\pi t},
$$

it follows that there exists a positive number t_0 such that

$$
e^{\pi t/2}\Psi(t) \le \sum_{n=1}^\infty e^{-(n^2-(1/2))\pi t_0}, \quad t \ge t_0.
$$

Thus,

$$\Psi(t) = O(e^{-\pi t/2})$$

as $t \to \infty$. So, the function

$$\mathbb{C} \ni s \mapsto \int_1^\infty \Psi(t)(t^{s/2} + t^{(1-s)/2})\frac{dt}{t} \in \mathbb{C}$$

is entire. Hence, the function $\pi^{-s/2}\Gamma\left(\frac{s}{2}\right)\zeta(s)$ is a holomorphic function on $\mathbb{C} \setminus \{0, 1\}$. But Γ has a simple pole at 0, so $\frac{1}{\Gamma}$ has a simple zero at 0. Therefore ζ can be holomorphically continued to a meromorphic function on $\mathbb{C}$ with only a simple pole at $s = 1$. Since

$$\lim_{s \to 1}(s-1)\zeta(s)$$

$$= \lim_{s \to 1}\pi^{s/2}\frac{1}{\Gamma\left(\frac{s}{2}\right)}\left(1 - \frac{s-1}{s} + (s-1)\int_1^\infty \Psi(t)(t^{s/2} + t^{(1-s)/2})\frac{dt}{t}\right) = 1,$$

we see that the residue of ζ at the simple pole $s = 1$ is equal to 1.

Finally, interchanging s and $1-s$ in (24.3), we obtain

$$\pi^{-s/2}\Gamma\left(\frac{s}{2}\right)\zeta(s) = \pi^{-(1-s)/2}\Gamma\left(\frac{1-s}{2}\right)\zeta(1-s)$$

for all $s \in \mathbb{C} \setminus \{0, 1\}$.

Remark 24.3 The contents in this chapter are abridged from [4].

Historical Notes

Hjalmar Mellin (1854–1933), a Finnish mathematician, is best remembered for the Mellin transform. He was educated at the University of Helsinki. He taught at the Polytechnic Institute in Helsinki that was later renamed Helsinki University of Technology where he was the first rector. The Helsinki University of Technology was merged with other institutes of higher education in Helsinki to become Aalto University in 2010.

Exercises

1. Let f be a C^∞ function on the unit circle $\mathbb{S}^1$ centered at the origin in $\mathbb{R}^2$. Use Fourier series in this chapter to find a solution u on $\mathbb{S}^1 \times [0, \infty)$ of the initial value problem for the heat equation

$$\begin{cases} \frac{\partial u}{\partial t}(z, t) = (\Delta_{\mathbb{S}^1} u)(z, t), & z \in \mathbb{S}^1, \ t > 0, \\ u(\cdot, 0) = f, \end{cases} \tag{24.4}$$

where $\Delta_{\mathbb{S}^1}$ is the Laplacian on $\mathbb{S}^1$. (Hint: Using polar coordinates in $\mathbb{R}^2$,

$$\Delta_{\mathbb{S}^1} = \frac{\partial^2}{\partial\theta^2}, \quad \theta \in [-\pi, \pi].)$$

2. Let t be a positive number. Find the heat kernel K_t of $\Delta_{\mathbb{S}^1}$, *i.e.*, K_t is the function on $\mathbb{S}^1$ such that

$$u(\theta, t) = \frac{1}{2\pi} \int_{-\pi}^{\pi} K_t(\theta, \phi) f(\phi) \, d\phi, \quad \theta \in [-\pi, \pi],$$

is a solution of the initial value problem (24.4).

3. Find a formula for the Green function G of $\Delta_{\mathbb{S}^1}$, *i.e.*, G is the function on $\mathbb{S}^1$ such that

$$u(\theta) = \frac{1}{2\pi} \int_{-\pi}^{\pi} G(\theta - \phi) f(\phi) \, d\phi, \quad \theta \in [-\pi, \pi],$$

is a solution of the partial differential equation

$$\Delta_{\mathbb{S}^1} u = f,$$

where f is a C^∞ function on $\mathbb{S}^1$.

4. Prove that the Riemann zeta-function ζ is holomorphic on the open right half plane $\{s \in \mathbb{C} : \operatorname{Re} s > 1\}$.

5. Use the well-known formula

$$\sum_{n=1}^{\infty} \frac{1}{n^2} = \frac{\pi^2}{6}$$

to prove that

$$\zeta(-1) = -\frac{1}{12}. \tag{24.5}$$

6. Prove that

$$\zeta(0) = -\frac{1}{2}. \tag{24.6}$$

7. Use a logarithmic derivative to prove that

$$\zeta'(0) = -\frac{1}{2}\ln(2\pi). \tag{24.7}$$

Chapter 25

The Twisted Bi-Laplacian

We begin with a recall from Chapter 17 that the twisted Laplacian L on $\mathbb{R}^2$ is the second-order partial differential operator given by

$$L = -\Delta + \frac{1}{4}(x^2 + y^2) - i\left(x\frac{\partial}{\partial y} - y\frac{\partial}{\partial x}\right), \tag{25.1}$$

where

$$\Delta = \frac{\partial^2}{\partial x^2} + \frac{\partial^2}{\partial y^2}.$$

Thus, the twisted Laplacian L is the Hermite operator

$$H = -\Delta + \frac{1}{4}(x^2 + y^2)$$

perturbed by the partial differential operator $-iN$, where

$$N = x\frac{\partial}{\partial y} - y\frac{\partial}{\partial x}$$

is the rotation operator.

By Theorem 21.5, the spectrum of the twisted Laplacian L is the set of all odd natural numbers. It should be noted, however, that each eigenvalue has infinite multiplicity.

Renormalizing the twisted Laplacian L to the partial differential operator P given by

$$P = \frac{1}{2}(L + 1), \tag{25.2}$$

we see that the eigenvalues of P are the natural numbers $1, 2, \ldots$, and each eigenvalue, as in the case of L, has infinite multiplicity.

Now, the conjugate $\overline{L}$ of the twisted Laplacian L is given by

$$\overline{L} = -\Delta + \frac{1}{4}(x^2 + y^2) + i\left(x\frac{\partial}{\partial y} - y\frac{\partial}{\partial x}\right) \tag{25.3}$$

and after renormalization, we get the conjugate Q of P given by

$$Q = \frac{1}{2}(\overline{L} + 1). \tag{25.4}$$

DOI: 10.1201/9781003206781-25

The aim of this chapter is to analyze the heat semigroup and the inverse of the twisted bi-Laplacian M defined by

$$M = QP = PQ = \frac{1}{4}(H - iN + 1)(H + iN + 1), \tag{25.5}$$

where P and Q commute because it can be shown by easy computations that H and N commute, *i.e.*, $HNf = NHf$ for all functions f in $C^\infty(\mathbb{R}^2)$.

Using Theorem 21.5 and the spectral mapping theorem, we have a complete spectral analysis of the twisted bi-Laplacian M given in the following theorem.

Theorem 25.1 *The eigenvalues and the eigenfunctions of the twisted bi-Laplacian M are, respectively, the natural numbers $1, 2, 3, \ldots$, and the functions $e_{j,k}$, $j, k = 0, 1, 2, \ldots$. More precisely, for $n = 1, 2, 3, \ldots$, the eigenfunctions corresponding to the eigenvalue n are all the functions $e_{j,k}$ where $j, k = 0, 1, 2, \ldots$, such that*

$$(j + 1)(k + 1) = n.$$

By means of Theorem 25.1, we see that the multiplicity of each eigenvalue n of the twisted bi-Laplacian is equal to the number $d(n)$ of divisors of the positive integer n. We give as Corollary 1.2 in [15] an estimate on the counting function $N(\lambda)$ defined as the number of eigenvalues of M less than or equal to λ. In fact, we can see that the following result, which is Corollary 1.2 in [15], is the well-known result on the asymptotic behavior of the counting function of the eigenvalues of the twisted bi-Laplacian, in which the multiplicity of each eigenvalue is taken into account.

Theorem 25.2 *For all λ in $[0, \infty)$,*

$$N(\lambda) = \sum_{n \le \lambda} d(n) = \lambda \ln \lambda + (2\gamma - 1)\lambda + E(\lambda), \tag{25.6}$$

where γ is Euler's constant and

$$E(\lambda) = O(\sqrt{\lambda})$$

as $\lambda \to \infty$.

Remark 25.3 Euler's constant γ is given by

$$\gamma = \lim_{m \to \infty} \left\{ \sum_{n=1}^{m} \frac{1}{n} - \ln m \right\} \approx 0.57721.$$

A complete and classical proof of Theorem 25.2 can be based on Theorem 3.12 in Chapter 8 of [43] and the above-mentioned connection between the divisors and the twisted bi-Laplacian. It is interesting to point out the connection with the Dirichlet divisor problem, which asks for the best number μ such that

$$E(\lambda) = O(\lambda^\mu)$$

as $\lambda \to \infty$. The conjecture is that $\mu = 1/4$, but it is a result of Hardy [18] that $\mu = 1/4$ does not work. Due to this celebrated connection, the positive divisors of a positive integer n are also called the Dirichlet divisors of n.

Another version of Theorem 25.2 with a different proof can be found at the end of this chapter.

We can use Theorem 25.2 to compute the trace $\mathrm{tr}(e^{-tM})$ of the heat semigroup generated by the twisted bi-Laplacian M, *i.e.*, the sum of the eigenvalues of e^{-tM}, where the multiplicity of each eigenvalue is counted.

Theorem 25.4 *For $t > 0$,*

$$\mathrm{tr}(e^{-tM}) = (\gamma - \ln t)t^{-1} + O(t^{\mu}),$$

where $\mu > \frac{1}{4}$.

Proof Since

$$\mathrm{tr}(e^{-tM}) = \int_0^{\infty} e^{-t\lambda} dN(\lambda),$$

it follows from an integration by parts that for $t > 0$,

$$\mathrm{tr}(e^{-tM}) = e^{-t\lambda} N(\lambda)\big|_0^{\infty} + t \int_0^{\infty} e^{-t\lambda} N(\lambda) d\lambda = t \int_0^{\infty} e^{-t\lambda} N(\lambda)\, d\lambda. \quad (25.7)$$

So, using (25.7) and the formula of $N(\lambda)$ in (25.6), we get for $t > 0$,

$$\begin{aligned}
\mathrm{tr}(e^{-tM}) &= t \int_0^{\infty} e^{-t\lambda}(\lambda \ln \lambda + (2\gamma - 1)\lambda + O(\lambda^{\mu}))\, d\lambda \\
&= t \int_0^{\infty} e^{-t\lambda} \lambda \ln \lambda\, d\lambda + (2\gamma - 1)t^{-1} + O(t^{\mu}). \quad (25.8)
\end{aligned}$$

Since

$$\begin{aligned}
\int_0^{\infty} e^{-t\lambda} \lambda \ln \lambda\, d\lambda &= -\frac{d}{dt} \int_0^{\infty} e^{-t\lambda} \ln \lambda\, d\lambda = \frac{d}{dt}\left[\frac{1}{t}(\gamma + \ln t)\right] \\
&= (1 - \gamma - \ln t)t^{-2}, \quad (25.9)
\end{aligned}$$

it follows from (25.8) and (25.9) that for $t > 0$,

$$\mathrm{tr}(e^{-tM}) = (\gamma - \ln t)t^{-1} + O(t^{\mu}),$$

as required. $\qquad\square$

We leave it as an exercise to show that the trace of the inverse M^{-1} of the twisted bi-Laplacian M does not exist. In other words, the sum of the eigenvalues of M^{-1} does not exist, but there is a version of the trace that is tailored for the inverse M^{-1} of the twisted bi-Laplacian M.

Let A be a positive and compact operator on a complex and separable Hilbert space X. Let

$$\lambda_1(A) \geq \lambda_2(A) \geq \cdots$$

be the eigenvalues of A arranged in decreasing order with multiplicities counted. For a positive integer k, we say that A is in the kth trace class if

$$\left\{ \frac{1}{\ln^k m} \sum_{j=1}^{m} \lambda_j(A) \right\}_{m=2}^{\infty} \in l^{\infty}.$$

If A is in the kth trace class such that

$$\lim_{m \to \infty} \frac{1}{\ln^k m} \sum_{j=1}^{m} \lambda_j(A)$$

exists, then the kth trace $\mathrm{tr}_k(A)$ of A is given by

$$\mathrm{tr}_k(A) = \lim_{m \to \infty} \frac{1}{\ln^k m} \sum_{j=1}^{m} \lambda_j(A).$$

Using Theorem 25.2, we get the following theorem for the trace of M^{-1}.

Theorem 25.5 *M^{-1} is in the second trace class and*

$$\mathrm{tr}_2(M^{-1}) = \frac{1}{2}.$$

Proof Let us compute $\sum_{n \leq x} \frac{d(n)}{n}$ for large and positive integers x, say, for $x > 2$. To do this, we use the partial summation formula to the effect that

$$\sum_{n \leq x} a_n f(n) = S(x-1)f(x) - \int_1^x S(t)f'(t)\, dt, \qquad (25.10)$$

where $\{a_n\}_{n=1}^{\infty}$ is a sequence with positive terms, f is a positive and differentiable function on $(0, \infty)$, and S is the function on $[1, \infty)$ given by

$$S(t) = \sum_{n \leq t} a_n, \qquad t \geq 1. \qquad (25.11)$$

Indeed,

$$
\begin{aligned}
\int_1^x S(t)f'(t)\, dt &= \sum_{n=1}^{x-1} \int_n^{n+1} S(t)f'(t)\, dt \\
&= \sum_{n=1}^{x-1} \int_n^{n+1} \left(\sum_{k=1}^{n} a_k \right) f'(t)\, dt \\
&= \sum_{n=1}^{x-1} \sum_{k=1}^{n} a_k(f(n+1) - f(n)).
\end{aligned}
$$

Interchanging the order of summation, we get

$$\int_1^x S(t)f'(t)\,dt = \sum_{k=1}^{x-1}\sum_{n=k}^{x-1} a_k(f(n+1) - f(n))$$

$$= \sum_{k=1}^{x-1} a_k(f(x) - f(k)).$$

Therefore

$$S(x-1)f(x) - \int_1^x S(t)f'(t)\,dt = \sum_{n=1}^x a_n f(n),$$

which is (25.10). Applying (25.10) and (25.11) with $a_n = d(n)$ and $f(n) = \frac{1}{n}$, and using the asymptotic formula for the function S as given by the Dirichlet divisor problem, we get

$$\sum_{n\leq x}\frac{d(n)}{n} = S(x-1)f(x) - \int_1^x S(t)f'(t)\,dt$$

$$= \frac{1}{x}((x-1)\ln(x-1) + (2\gamma-1)(x-1) + O(\sqrt{x}))$$

$$+ \int_1^x \left(\frac{\ln t}{t} + (2\gamma-1)t^{-1} + O(t^{-3/2})\right)\,dt. \qquad (25.12)$$

Since

$$(x-1)\ln(x-1) = x\ln x + O(\sqrt{x}) \qquad (25.13)$$

as $x \to \infty$, and

$$\int_1^x \frac{\ln t}{t}\,dt = \frac{1}{2}\ln^2 x. \qquad (25.14)$$

it follows from (25.12)–(25.14) that

$$\sum_{n\leq x}\frac{d(n)}{n} = \frac{1}{x}(x\ln x + (2\gamma-1)x + O(\sqrt{x}))$$

$$+ \frac{1}{2}\ln^2 x + (2\gamma-1)\ln x + O(x^{-1/2})$$

$$= \frac{1}{2}\ln^2 x + 2\gamma \ln x + (2\gamma-1) + O(x^{-1/2})$$

as $x \to \infty$. This completes the proof. $\qquad\Box$

We now give the following meromorphic extension of the Riemann zeta-function ζ_M for the twisted bi-Laplacian M from $\{s \in \mathbb{C} : \operatorname{Re} s > 1\}$ to $\mathbb{C}\setminus\{1\}$ with only a double pole at $s = 1$.

Theorem 25.6 ζ_M, *initially defined on* $\{s \in \mathbb{C} : \operatorname{Re} s > 1\}$, *can be extended to a meromorphic function on* $\mathbb{C}$ *with only a double pole at* $s = 1$ *and the Laurent series of* ζ_M *centered at 1 is given by*

$$\zeta_M(s) = \frac{1}{(s-1)^2} + \frac{2\gamma}{s-1} + \text{the holomorphic part}, \quad s \in \mathbb{C} \setminus \{1\}.$$

Proof From Chapter 24, we know that the Riemann zeta-function ζ, initially defined on $\{z \in \mathbb{C} : \operatorname{Re} s > 1\}$, can be extended to a meromorphic function on $\mathbb{C}$ with the only simple pole at $s = 1$. By the result in [5], the Laurent series of ζ centered at $s = 1$ is given by

$$\zeta(s) = \frac{1}{s-1} + \sum_{n=0}^{\infty} \frac{(-1)^n}{n!} \gamma_n (s-1)^n, \quad s \in \mathbb{C} \setminus \{1\},$$

where $\gamma_0, \gamma_1, \gamma_2, \ldots$, are the Stieltjes constants given by

$$\gamma_n = \lim_{m \to \infty} \left[\left(\sum_{k=1}^{m} \frac{\ln^n k}{k} \right) - \frac{\ln^{n+1} m}{n+1} \right], \quad n = 0, 1, 2, \ldots.$$

We note that in particular

$$\gamma_0 = \lim_{m \to \infty} \left[\left(\sum_{k=1}^{m} \frac{1}{k} \right) - \ln m \right],$$

which is Euler's constant γ in Chapter 25. By Lemma 26.1 and multiplication of the Laurent series of ζ by itself, we get

$$\zeta_M(s) = \zeta^2(s) = \frac{1}{(s-1)^2} + \frac{2\gamma}{s-1} + \text{the holomorphic part} \qquad (25.15)$$

for all $s \in \mathbb{C} \setminus \{1\}$. $\qquad\qquad\square$

We need some very basic functional analysis in order to find an asymptotic expansion of the counting function for the eigenvalues of the twisted bi-Lapacian M. Let A be a linear operator from a complex and separable Hilbert space X into X such that A has positive eigenvalues with finite multiplicities arranged in increasing order

$$\lambda_1(A) \le \lambda_2(A) \le \cdots \to \infty$$

and the multiplicity of each eigenvalue is counted. Suppose that the Riemann zeta-function of A given by

$$\zeta_A(s) = \operatorname{tr}(A^{-s})$$

can be extended to a meromorphic function on $\mathbb{C}$ with only a pole of order p at $s = a$ and the Laurent series of ζ_A centered at a is given by

$$\zeta_A(s) = \sum_{j=1}^{p} \frac{a_{-j}}{(s-a)^j} + \text{the holomorphic part}, \quad s \in \setminus \{a\}.$$

Then, according to the theorem on page 15 of [2], there exists a positive number δ such that as $\lambda \to \infty$,

$$N_A(\lambda) = \sum_{j=1}^{p} \frac{a_{-j}}{(j-1)!} \left(\frac{d}{ds}\right)^{j-1} \Bigg|_{s=a} \left(\frac{\lambda^s}{s}\right) + O(\lambda^{a-\delta}). \tag{25.16}$$

We end this chapter with another asymptotic expansion of the eigenvalue counting function of the twisted bi-Laplacian M.

Theorem 25.7 *Let $N(\lambda)$ denote the number of eigenvalues of M less than or equal to λ. Then there exists a positive number δ such that*

$$N(\lambda) = \sum_{n \leq \lambda} d(n) = \lambda \ln \lambda + (2\gamma - 1)\lambda + E(\lambda),$$

where

$$E(\lambda) = O(\lambda^{1-\delta})$$

as $\lambda \to \infty$.

Proof By (25.15) and (25.16), there exists a positive number δ such that

$$
\begin{aligned}
N(\lambda) &= 2\gamma \frac{\lambda^s}{s}\Bigg|_{s=1} + \frac{d}{ds}\Bigg|_{s=1} \left(\frac{\lambda^s}{s}\right) + O(\lambda^{1-\delta}) \\
&= 2\gamma\lambda + \frac{s\lambda^s \ln \lambda - \lambda^s}{s^2}\Bigg|_{s=1} + O(\lambda^{1-\delta}) \\
&= 2\gamma\lambda + \lambda \ln \lambda - \lambda + O(\lambda^{1-\delta}) \\
&= \lambda \ln \lambda + (2\gamma - 1)\lambda + O(\lambda^{1-\delta})
\end{aligned}
$$

as $\lambda \to \infty$. $\qquad\square$

Historical Notes

Godfrey Harold Hardy (1877–1947) was an English mathematician educated at the University of Cambridge. He taught at the University of Cambridge from 1906 to 1919. He then left Cambridge to become Savilian Professor of Geometry at the University of Oxford and returned to be Sadleirian Professor of Pure Mathematics at the University of Cambridge in 1931. He was the first mathematician who introduced rigor into the teaching of calculus in England as manifested in his book "A Course of Pure Mathematics" first published by Cambridge University Press in 1908. The book aged well to the tenth edition in 1952. Many reprintings of the tenth edition are still good sellers up to now. He was a world-class expert in analysis and number theory. In addition, he was an excellent writer. His nontechnical and popular book "A Mathematician's Apology", also published by Cambridge University Press, is an ideal book about mathematics and mathematicians for the general public.

Thomas Joannes Stieltjes (1856–1894) was a Dutch mathematician. He worked in the moment problem, number theory, and continued fractions, among others. He is best remembered for the Riemann–Stieltjes integral, which is studied by undergraduate students in analysis nowadays.

Exercises

1. Let H and N be partial differential operators on $\mathbb{R}^2$ given by

$$H = -\Delta + \frac{1}{4}(x^2 + y^2)$$

and

$$N = x\frac{\partial}{\partial y} - y\frac{\partial}{\partial x}. \tag{25.17}$$

Prove that for all $f \in C^2(\mathbb{R}^2)$,

$$HNf = NHf.$$

2. Prove that the Dirichlet divisor function is multiplicative , *i.e.*,

$$d(mn) = d(m)d(n)$$

for all positive integers m and n with $(m, n) = 1$. (The formula

$$(m, n) = 1$$

means that the greatest common divisor of m and n is 1.)

3. Let N be as in (25.17). Find $\text{tr}(e^{-tN})$ by first expressing N in terms of polar coordinates in $\mathbb{R}^2$.

4. Prove that the trace $\text{tr}(M^{-1})$ of the inverse M^{-1} of the twisted bi-Laplacian M does not exist.

Chapter 26

Complex Powers of the Twisted Bi-Laplacian

Let $\alpha \in \mathbb{C}$. Then we can define the complex power M^α of the twisted bi-Laplacian M in terms of the eigenvalues $\{n^\alpha : n = 1, 2, \ldots\}$ with multiplicities counted. Of course, the principal branch of n^α for each positive integer n is used.

Let $\alpha \in \mathbb{C}$. Then we define the complex-valued function ζ_{M^α} on $\mathbb{C}$ *formally* by

$$\zeta_{M^\alpha}(s) = \operatorname{tr}((M^\alpha)^{-s}) = \operatorname{tr}(M^{-\alpha s}), \quad s \in \mathbb{C}.$$

We begin with the following easy observation.

Lemma 26.1 *Let $\alpha \in \mathbb{C}$. Then for all s in $\mathbb{C}$ with $\operatorname{Re}(\alpha s) > 1$,*

$$\zeta_{M^\alpha}(s) = \zeta^2(\alpha s).$$

Proof Let $s \in \mathbb{C}$ be such that $\operatorname{Re}(\alpha s) > 1$. Then by Theorem 25.1, the eigenvalues of $M^{-\alpha s}$ are $n^{-\alpha s}$, $n = 1, 2, \ldots$, and the multiplicity of $n^{-\alpha s}$ is equal to the number $d(n)$ of Dirichlet divisors of n. Therefore

$$\zeta_{M^\alpha}(s) = \sum_{n=1}^{\infty} \frac{d(n)}{n^{\alpha s}}. \tag{26.1}$$

So, a straightforward computation gives

$$\zeta^2(\alpha s) = \sum_{\mu=1}^{\infty} \frac{1}{\mu^{\alpha s}} \sum_{\nu=1}^{\infty} \frac{1}{\nu^{\alpha s}} = \sum_{n=1}^{\infty} \frac{1}{n^{\alpha s}} \sum_{\mu\nu=n} 1 = \sum_{n=1}^{\infty} \frac{d(n)}{n^{\alpha s}}.$$

$\square$

The Riemann zeta-function regularizations of the trace and the determinant of M^α, denoted by $\operatorname{tr}_R(M^\alpha)$ and $\det_R(M^\alpha)$, respectively, are defined by

$$\operatorname{tr}_R(M^\alpha) = \zeta_{M^\alpha}(-1)$$

and

$$\det_R(M^\alpha) = e^{-\zeta'_{M^\alpha}(0)}.$$

The meanings of these two quantities can be found in, for instance, the book [30].

DOI: 10.1201/9781003206781-26

Theorem 26.2 *Let $\alpha \in \mathbb{C} \setminus \{-1\}$. Then*

$$\mathrm{tr}_R(M^\alpha) = \zeta^2(-\alpha).$$

Proof By Lemma 26.1 and the holomorphic continuation of the Riemann zeta-function to a meromorphic function on $\mathbb{C}$ with only a simple pole at $s = 1$, we see that

$$\mathrm{tr}_R(M^\alpha) = \zeta_{M^\alpha}(-1) = \zeta^2(-\alpha).$$

□

Remark 26.3 By (24.5),

$$\zeta(-1) = -\frac{1}{12}.$$

Hence

$$\mathrm{tr}_R(M) = \frac{1}{144}.$$

Remark 26.4 In the case when $\alpha = -1$, the Riemann zeta-function regularization of the trace of the inverse M^{-1} is equal to infinity. The 2nd trace instead of the Riemann zeta-function regularization of the trace of the inverse M^{-1} is a finite number.

Theorem 26.5 *Let $\alpha \in \mathbb{C}$. Then*

$$\det_R(M^\alpha) = (2\pi)^{-\alpha/2}.$$

Proof As in Theorem 26.2,

$$\det_R(M^\alpha) = e^{-\zeta'_{M^\alpha}(0)} = e^{-2\alpha\zeta(0)\zeta'(0)}.$$

By (24.6), $\zeta(0) = -\frac{1}{2}$ and by (24.7), $\zeta'(0) = -\frac{1}{2}\ln(2\pi)$. So,

$$\det_R(M^\alpha) = (2\pi)^{-\alpha/2}.$$

□

As an application, we can give a formula for the Riemann zeta-function regularizations of the determinants of the heat semigroups of complex powers of the twisted bi-Laplacian.

Theorem 26.6 *Let $\alpha \in \mathbb{C} \setminus \{-1\}$. Then for $t > 0$,*

$$\det_R(e^{-tM^\alpha}) = e^{-t\zeta^2(-\alpha)}.$$

Proof By Theorem 25.1 and the spectral mapping theorem, the eigenvalues of $e^{(-tM^\alpha)^{-s}}$ are $e^{tn^\alpha s}$, $n = 1, 2, \ldots$, and the multiplicity of the eigenvalue $e^{tn^\alpha s}$ is $d(n)$. Therefore

$$\zeta_{e^{-tM^\alpha}}(s) = \mathrm{tr}((e^{-tM^\alpha})^{-s}) = \sum_{n=1}^{\infty} d(n)e^{tn^\alpha s}, \quad s \in \mathbb{C}.$$

So, by (26.1) and Theorem 26.2,

$$\zeta'_{e^{-tM^\alpha}}(0) = t \sum_{n=1}^{\infty} d(n)n^\alpha = t\zeta^2(-\alpha).$$

Thus,

$$\det{}_R(e^{-tM^\alpha}) = e^{-\zeta'_{e^{-tM^\alpha}}(0)} = e^{-t\zeta^2(-\alpha)},$$

and this completes the proof. □

Remark 26.7 By Theorems 26.2 and 26.6, we see that for $\alpha \in \mathbb{C} \setminus \{-1\}$,

$$\det{}_R(e^{-tM^\alpha}) = e^{-t(\mathrm{tr}_R(M^\alpha))}, \quad t > 0,$$

which is in conformity with the well-known relationship between the determinant and the trace of a square matrix A given by

$$\det(e^A) = e^{\mathrm{tr}(A)}.$$

Exercises

1. Prove that the Reimann zeta-function regularization of the trace of the inverse M^{-1} of the twisted bi-Laplacian M does not exist.

2. Does $\mathrm{tr}_R(M^{-2})$ exist? If yes, compute its value. If no, explain why not.

3. Find all complex numbers α for which $\mathrm{tr}_R(M^\alpha) = 0$ assuming that the Riemann hypothesis is true. (The Riemann hypothesis states that all the zeros in $\mathbb{C} \setminus \mathbb{R}$ form a countably infinite subset of the vertical line $\{s \in \mathbb{C} : \mathrm{Re}\, s = \frac{1}{2}\}$.)

Bibliography

[1] S. Agmon, *Lectures on Elliptic Boundary Value Problems*, AMS Chelsea Publication, Providence, RI, 2010.

[2] J. Aramaki, On an extension of Ikehara's Tauberian theorem, *Pacific J. Math.* **133** (1988), 13–30.

[3] E. Artin, *The Gamma Function*, Holt, Reinhart and Winston, New York, 1964.

[4] R. Bellman, *A Brief Introduction to Theta Functions*, Holt, Reinhart and Winston, New York, 1961.

[5] W. E. Briggs, Some constants associated with the Riemann zeta-function, *Michigan Math. J.* **3** (1955/56), 117–121.

[6] X. Duan, *The Heat Kernel and Green Function of the Sub-Laplacian on the Heisenberg Group*, M.A. Survey Paper, York University, Toronto, ON, 2012.

[7] X. Duan, *Complex Powers of a Fourth-Order Operator: Heat Kernels, Green Functions and L^p-$L^{p'}$ Estimates*, Ph.D. Dissertation, York University, Toronto, ON, 2016.

[8] X. Duan and M. W. Wong, The Dirichlet divisor problem, traces and determinants for complex powers of the twisted bi-Laplacian, *J. Math. Anal. Appl.* **410** (2014), 151–157.

[9] G. B. Folland, A fundamental solution for a subelliptic operator, *Bull. Amer. Math. Soc.* **79** (1973), 373–376.

[10] G. B. Folland, *Harmonic Analysis in Phase Space*, Princeton University Press, Princeton, NJ, 1989.

[11] A. Friedman, *Partial Differential Equations*, Holt, Reinhart and Winston, New York, 1969.

[12] A. Friedman, *Foundations of Modern Analysis*, Holt, Reinhart and Winston, New York, 1970.

[13] B. Gaveau, Principe de moindre action, propagation de la chaleur, et etsimées sous, elliptiques sur certains groupes nilpotents, *Acta Math.* **139** (1977), 95–153.

[14] R. R. Goldberg, *Fourier Transforms*, Cambridge University Press, Cambridge, 1961.

[15] T. Gramchev, S. Pilipović, L. Rodino, and M. W. Wong, Spectral properties of the twisted bi-Laplacian, *Arch. Math.* **93** (2009), 565–575.

[16] L. Greengard and J. Stark, The fast Gauss transform, *SIAM J. Sci. Stat. Comput.* **12** (1991), 79–94.

[17] A. Grossmann, G. Loupias, and E. M. Stein, An algebra of pseudodifferential operators and quantum mechanics in phase space, *Ann. Inst. Fourier (Grenoble)* **18** (1968), 343–368.

[18] G. H. Hardy, On Dirichlet divisor problem, *Proc. Lond. Math. Soc.* **15** (1916), 1–25.

[19] G. H. Hardy, *Divergent Series*, AMS Chelsea Publication, Providence, RI, 2000.

[20] L. Hörmander, On the theory of general linear partial differential operators, *Acta Math.* **94** (1955), 161–248.

[21] L. Hörmander, Hypoelliptic second-order differential equations, *Acta Math.* **119** (1967), 147–171.

[22] L. Hörmander, *Linear Partial Differential Operators*, Springer-Verlag, Berlin, 1976.

[23] L. Hörmander, *The Analysis of Linear Partial Differential Operators I: Distribution Theory and Fourier Analysis*, Reprint of the Second (1990) Edition, Classics in Mathematics, Springer-Verlag, Berlin, 2003.

[24] L. Hörmander, *The Analysis of Linear Partial Differential Operators II: Differential Operators with Constant Coefficients*, Reprint of the 1983 Original, Classics in Mathematics, Springer-Verlag, Berlin, 2005.

[25] L. Hörmander, *The Analysis of Linear Partial Differential Operators III: Pseudo-Differential Operators*, Reprint of the 1994 Edition, Classics in Mathematics, Springer-Verlag, Berlin, 2007.

[26] L. Hörmander, *The Analysis of Linear Partial Differential Operators IV: Fourier Integral Operators*, Reprint of the 1994 Edition, Classics in Mathematics, Springer-Verlag, Berlin, 2009.

[27] A. Hulanicki, The distribution of energy in the Brownian motion in the Gaussian field and analytic hypoellipticity of certain subelliptic operators on the Heisenberg group, *Studia Math.* **56** (1976), 165–173.

[28] S. G. Krantz, *Explorations in Harmonic Analysis with Applications to Complex Function Theory and the Heisenberg Group*, Birkhäuser, Boston, 2009.

[29] N. Levinson and R. M. Redheffer, *Complex Variables*, Holden-Day, San Francisco, 1970.

[30] S. Lord, F. Sukochev, and D. Zanin, *Singular Traces: Theory and Applications*, de Gruyter, Berlin, 2013.

[31] W. Magnus, F. Oberhettinger, and R. P. Soni, *Formulas and Theorems for the Special Functions of Mathematical Physics*, Springer-Verlag, Berlin, 1964

[32] J. E. Moyal, Quantum mechanics as a statistical theory, *Proc. Camb. Phil. Soc.* **45** (1949), 99–124.

[33] M. Reed and B. Simon, *Fourier Analysis, Self-Adjointness*, Academic Press, Waltham, MA, 1975.

[34] M. Reed and B. Simon, *Functional Analysis,* Revised and Large Edition, Academic Press, Waltham, MA, 1980.

[35] M. Schechter, *Modern Methods in Partial Differential Equations, An Introduction*, McGraw-Hill, New York, 1977.

[36] M. Schechter, *Spectra of Partial Differential Operators*, Second Edition, North-Holland, Amsterdam, 1986.

[37] M. Schechter, *Principles of Functional Analysis*, Second Edition, American Mathematical Society, Providence, RI, 2002.

[38] M. Schechter, *Solving Linear Partial Differential Equations: Spectra*, World Scientific, Singapore, 2021.

[39] M. Shubin, *Pseudodifferential Operators and Spectral Theory*, Springer-Verlag, Berlin, 1978.

[40] B. Simon, Distributions and their Hermite expansions, *J. Math. Phys.* **12** (1971), 140–148.

[41] E. M. Stein, *Singular Integrals and Differentiability Properties of Functions,* Princeton University Press, Princeton, NJ, 1970.

[42] E. M. Stein, *Harmonic Analysis: Real-variable Methods, Orthogonality, and Oscillatory Integral*, Princeton University Press, Princeton, NJ, 1993.

[43] E. M. Stein and R. Shakarchi, *Fourier Analysis: An Introduction*, Princeton University Press, NJ, 2003.

[44] E. M. Stein and R. Shakarchi, *Complex Analysis*, Princeton University Press, Princeton, NJ, 2003.

[45] E. M. Stein and G. Weiss, *Introduction to Fourier Analysis on Euclidean Spaces,* Princeton University Press, Princeton, NJ, 1971.

[46] R. S. Strichartz, *A Guide to Distribution Theory and Fourier Transforms,* CRC Press, Boca Raton, FL, 1994.

[47] S. Thangavelu, *Lectures on Hermite and Laguerre Expansions,* Princeton University Press, Princeton, NJ, 1993.

[48] S. Thangavelu, *Harmonic Analysis on the Heisenberg Group*, Birkhäuser, Boston, 1998.

[49] E. C. Titchmarsh, *The Theory of the Riemann-Zeta Function, Second Edition,* Oxford University Press, 1986, with a preface by D. R. Heath-Brown.

[50] F. Treves, *Basic Linear Partial Differential Equations*, Academic Press, Waltham, MA, 1975.

[51] H. F. Weinberger, *A First Course in Partial Differential Equations with Complex Variables and Transform Methods,* Blaisedel, New York, 1965.

[52] H. Weyl, *The Theory of Groups and Quantum Mechanics*, Dover, Mineola, NY, 1950.

[53] E. Wigner, On the quantum correction for thermodynamic equilibrium, *Phys. Rev.* **40** (1932), 749–759.

[54] M. W. Wong, *Weyl Transforms,* Springer-Verlag, Berlin, 1998.

[55] M. W. Wong, Weyl transforms, the heat kernel and Green function of a degenerate elliptic operator, *Ann. Global Anal. Geom.* **28** (2005), 271–283.

[56] M. W. Wong, *Complex Analysis*, World Scientific, Singapore, 2008.

[57] M. W. Wong, *Discrete Fourier Analysis*, Birkhäuser, Basel, 2011.

[58] M. W. Wong, *An Introduction to Pseudo-Differential Operators,* Third Edition, World Scientific, Singapore, 2014.

[59] S. Yang, *Laplacians on Non-Isotropic Heisenberg Groups with Multi-Dimensional Center*, Ph.D. Dissertation, York University, Toronto, ON, 2021.

Index

Printed in the United States
by Baker & Taylor Publisher Services